Cucurbit Genetics Cooperative

31-32

2008-2009

ISSN 1064-5594

2102 Plant Sciences Building
College Park, Maryland
20742-4452 USA

Tel: (301) 405-1321 or (580) 889-7395
Fax: (301) 314-9308

binkley@umd.edu
angela.davis@lane-ag.org

http://cuke.hort.ncsu.edu/cgc/
http://www.umresearch.umd.edu/cgc/

The **Cucurbit Genetics Cooperative** (CGC) was organized in 1977 to develop and advance the genetics of economically important cucurbits. Membership to CGC is voluntary and open to individuals who have an interest in cucurbit genetics and breeding. CGC membership is on a biennial basis. For more information on CGC and its membership rates, visit our website (*http://cuke.hort.ncsu.edu/cgc)*

or contact Tim Ng, (301) 405-1321, *binkley@umd.edu*, or Angela Davis, (580)889-7395, angela.davis@lane-ag.org).

CGC Reports are issued on an annual basis. The Reports include articles submitted by CGC members for the use of CGC members. None of the information in the annual report may be used in publications without the consent of the respective authors for a period of five years.

ISSN 1064-5594

Contents

News & Comments

Cucurbit Genetics Cooperative Report Call for Papers

The call for papers for **CGC 33** (2010) is open, and we are **accepting papers** for the volume now. Send manuscripts to the appropriate crop editor. (*http://cuke.hort.ncsu.edu/cgc*) If you do not receive your copy, contact Linda Wessel-Beaver.

Comments from CGC Gene List Committee

List of known genes for the Cucurbitaceae have been published previously in Hortscience and in reports of the Cucurbit Genetics Cooperative. CGC is currently publishing complete lists of known genes for cucumber (*Cucumis sativus*), melon (*Cucumis melo*), watermelon (*Citrullus lanatus*), and *Cucurbita* spp. on a rotating basis.

It is hoped that scientists will consult these lists as well as the rules of gene nomenclature for the Cucurbitaceae before choosing a gene name and symbol. Thus, inadvertent duplication of gene names and symbols will be prevented. The rules of gene nomenclature were adopted in order to provide guidelines for the naming and symbolizing of genes previously reported and those which will be reported in the future. Scientists are urged to contact members of the Gene List Committee regarding questions in interpreting the nomenclature rules and in naming and symbolizing new genes.

- Cucumber: Yiqun Weng (curator) and Todd C. Wehner (assistant)
- Melon: Catherine Dogimont (curator) Michael Pitrat (assistant curator) and James D. McCreight (assistant curator)
- Other Genera: Mark G. Hutton (curator) and Thomas Andres (assistant curator)
- *Cucurbita* spp.: Harry Paris (curator) and Eileen Kabelka (assistant curator)
- Watermelon: Todd C. Wehner (curator) and Stephen R. King (assistant curator)

Comments from the CGC Gene Curators

CGC has appointed Curators for the four major cultivated groups: cucumber, melon, watermelon and *Cucurbita* spp.

Curators are responsible for collecting, maintaining and distributing upon request stocks of the known marker genes. CGC members are requested to forward samples of currently held gene stocks to the respective Curator.

- Cucumber: Yiqun Weng (curator) and Todd C. Wehner (assistant)
- Melon: Catherine Dogimont (curator) Michael Pitrat (assistant curator) and James D. McCreight (assistant curator)
- Other Genera: Mark G. Hutton (curator) and Thomas Andres (assistant curator)
- *Cucurbita* spp.: Harry Paris (curator) and Eileen Kabelka (assistant curator)
- Watermelon: Todd C. Wehner (curator) and Stephen R. King (assistant curator)

2009 Watermelon Research and Development Group – 29th Annual Meeting

By Elisabetta Vivoda

The Annual Meeting of the Watermelon Research & Development Working Group was held Sunday, February 1, 2009 at the Westin Peachtree Plaza in Atlanta, GA, from 8:00 a.m. to 5:00 p.m. The meeting was held in conjunction with The Southern Association of Agricultural Scientists and the Southern Region American Society for Horticultural Sciences (SR-ASHS). Following a welcome from Stephen King, reports from the following seed companies were given: Harris Moran (Brenda Lanini), Willhite, Abbott & Cobb, Syngenta (James Brusca), Zeraim Gedera and Sakata.

Statewide watermelon trial reports for 2007 were given by Gilbert Miller: Seedless and Miniwatermelon Variety Trials at EREC, 2002-2008; Jonathan Schultheis: 2008 North Carolina Variety Trials; George Boyhan: 2008 Georgia Variety Trials and 2007-08 Pollenizer trials; Juan Anciso: details on the web at http://aggie-horticulture.tamu.edu/vegetable/varietytrials/

After the trial results the following research reports were presented:

- **An economic evaluation of using watermelon juice in ethanol production.** Merritt J. Taylor*1, Wayne Fish2, Benny Bruton2 and Vince Russo2. 1Wes Watkins Agricultural Research & Extension Center, Oklahoma State University, Lane, OK, 2USDA-ARS, SCARL, Lane, OK. *(mtaylor-okstate@lane-ag.org).
- **Activities of the National Watermelon Promotion Board.** Mark Arney, Executive Director, NWPB.
- **Regional watermelon grafting effort evaluating effects on yield and quality in marketable melons.** Richard L. Hassell*1, Jonathan R. Schultheis2, Stephen M. Olson3, and William Terry Kelley4. 1Clemson University CREC, 2N.C. State, 3University of Florida, and 4University of Georgia. *(rhassel@clemson.edu).
- **Reaction of watermelon rootstocks to root-knot nematode in field tests.** J. A. Thies*1, J. J. Ariss1, R. L.

Hassell2, S. Olson3. 1U.S. Vegetable Laboratory, USDA, ARS, Charleston, SC; 2Coastal Research and Education Center, Clemson University, Charleston, SC; 3University of Florida, Quincy, FL. *(Judy.Thies@ARS.USDA.GOV).

- **Managing Fusarium wilt of watermelon with acibenzolar-S-methyl and fungicides.** K. L. Everts*1,2, and X. G. Zhou1, 1University of Maryland College Park, Salisbury and 2University of Delaware, Newark. *(keverts@umd.edu).
- **Inheritance of resistance to powdery mildew, a new disease of watermelon.** A.Y. Tetteh and T.C. Wehner*. North Carolina State University, Raleigh, NC. *(todd_wehner@ncsu.edu). **New fungicides for managing Phytophthora fruit rot of watermelon.** C.S. Kousik1 and R. Hassell2. 1USDA-ARS, US Vegetable Lab, 2700 Savannah Highway, Charleston, SC; 2Coastal Research and Education Center, Clemson University, Charleston, SC. *(Shaker.Kousik@ars.usda.gov).
- **Inheritance of high fruit yield in two watermelon populations.** R. Kumar and T.C. Wehner*. North Carolina State University, Raleigh, NC. *(todd_wehner@ncsu.edu).
- **Discovery of Markers Linked to the ZYMV-FL Resistance Gene and Their Use in Marker-assisted Selection in Watermelon.** Karen Harris1, Kai-Shu Ling1, William P. Wechter1, Amnon Levi1*, D.F. Jenelle2, Michael J. Havey2, Nihat Guner3 and Todd C. Wehner3. 1USDA-ARS, US Vegetable Lab, 2700 Savannah Highway, Charleston, SC; 2USDA-ARS, University of Wisconsin, Madison, WI; 3North Carolina State University, Raleigh, NC. *(Amnon.Levi@ars.usda.gov)
- **New flesh colors in watermelon?** Stephen R. King1*, Angela R. Davis2, and Haejeen Bang1. 1Texas A&M University, College Station, TX; 2USDA-ARS, SCARL, Lane, OK. *(srking@tamu.edu).

Topics for general discussion were:

- o Sizing problem on second crop watermelons in California. It was suggested that the development of secondary roots occurred by adding soil at the crown and adding fertilizers can mitigate the problem.
- o Fruit softening around seed cavity in seedless watermelons in Australia. The problem is most likely environmental, caused by water stress during fruit development and it is linked to two varieties.
- o Priority list for research topics. Addition of watermelon genome, molecular markers, research on citrulline and lycopene, P capsici resistance.

Elisabetta Vivoda assumed the position of chair of the group, replacing Stephen King.

Stephen contribution to the group and watermelon research is greatly appreciated.

Jonathan Schultheis was elected vice-chair.

Todd Wehner was re elected secretary.

The WRDG thanks Abbott & Cobb for providing refreshments at the meeting.

Comment from the U.S. Cucurbit Crop Germplasm Committee Chair

James D. McCreight

This group operates under the auspices of the USDA-ARS National Plant Germplasm System (NPGS), is composed of ARS, university and industry scientists, and provides guidance to NPGS on matters relating to cucurbit crops and wild related species. Committee membership and species-specific crop reports are accessible through the NPGS website: (http://www.ars-grin.gov/npgs/). The committee receives, reviews, and recommends germplasm evaluation proposals annually for funding by NPGS, and also reviews and recommends proposals for germplasm collection and exchange. Contact James D. McCreight, USDA-ARS, Salinas, Calif., U.S.A., james.mccreight@ars.usda.gov for more information.

Upcoming Meetings of Interest to Cucurbit Researchers

Cucurbitaceae 2012

Dear Colleagues,

We invite you to attend Cucurbitaceae 2010 to be held November 14-18 in Charleston, South Carolina, USA. The conference will be held at the historic Francis Marion Hotel in downtown Charleston. The goal of this conference is to bring together colleagues working with cucurbits so we can share information on every aspect of cucurbit research, development, and production. Program topics include Breeding and Genetics, Economics, Entomology, Growth and Development, Marketing, Metabolomics, New Technologies, Pathology, Physiology, Production, and Utilization and Processing. On Sunday, November 14, we are planning a special symposium on Cucurbit Rootstocks and Grafting. We welcome you to join us for this exciting and in-depth conference exploring new frontiers of cucurbit research and development. We look forward to seeing you in the beautiful, historic city of Charleston, South Carolina in November!

You may access the conference website at www.ashs.org/cucurbit2010

Conference contact: Judy Thies
judy.thies@ars.usda.gov
Tel: 843-402-5317 Fax: 843-573-4715

Judy Thies, Chair
Shaker Kousik, Amnon Levi, Conference Organizers

Upcoming Meetings & News of Interest

Organization/Meeting	*Dates*	*Location*	*Contact*
30th Annual Meeting of the Watermelon Research & Development Group	February 7, 2010 8:00 am -5:00 pm November 17th, 2010 5:30-6:30 pm	In conjunction with the 70th Annual Meeting of the Southern Region - American Society for Horticultural Science, Orlando, FL, USA In conjunction with Cucurbitaceae 2010, Charleston, SC, USA.	Elisabetta Vivoda E.Vivoda@hmclause.com Jonathan Schultheis jonathan_schultheis@ncsu.edu
ISHS Cucurbit Conference	TBA	TBA	TBA
Cucurbit Crop Germplasm Committee Meeting	November 16, 2010 5:30-7:30 pm	In conjunction with Cucurbitaceae 2010, Charleston, SC, USA.	Jim McCreight jmccreight@pw.ars.usda.gov
Cucurbit Genetics Cooperative Business Meeting	November 15, 2010 6:30-7:30 pm	In conjunction with Cucurbitaceae 2010, Charleston, SC, USA.	Todd Wehner todd_wehner@ncsu.edu
Pickle Packers International	April 13-15 October 19-21, 2010	Loews Philadelphia, Pennsylvania, USA. Hyatt Regency, San Antonio, Texas, USA.	Susan Fuller, PPI Program Associate 202-331-2466 http://www.ilovepickles.org
Cucurbitaceae 2010	November 14-18, 2010	Francis Marion Hotel, Charleston, SC, USA.	Conference Organizers, Judy Thies, Chair Shaker Kousik, Amnon Levi judy.thies@ars.usda.gov Tel: 843-402-5317 Fax: 843-573-4715
X EUCARPIA International Meeting on Cucurbitaceae Eucarpia 2012	TBA	Turkey	Nebahat Sari nesari@cu.edu.tr
Melon Breeders Group	November 17, 2010 6:30-7:30 pm	In conjunction with Cucurbitaceae 2010, Charleston, SC, USA.	TBA
National Watermelon Association	February 17-20, 2010 2011	Fairmont Dallas Hotel, Dallas, TX, USA San Diego, CA	Tel: 813-754-7575 Fax: 813-754-1118 nwa@tampabay.rr.com http://www.nationalwatermelonassociation.com
Squash Research Group	November 16, 2010 5:30-6:30 pm	In conjunction with Cucurbitaceae 2010, Charleston, SC, USA.	TBA
Pickling Cucumber Improvement Committee	November 15, 2010 5:30-6:30	In conjunction with Cucurbitaceae 2010, Charleston, SC, USA.	Yiqun Weng weng4@wisc.edu

Cucurbit Genetics Cooperative

Style Guide

The following *guidelines* are for use in the preparation of reports. It is recognized that CGC members may not be able to meet one or more of the guidelines.

Authors are encouraged to contribute reports even though some of the guidelines cannot be met.

Our objective is to facilitate the interchange of information, but we ask authors to help reduce unnecessary editing.

Refer to the latest Cucurbit Genetics Cooperative Report regarding questions of style not mentioned.

I. **Reports will be assigned to one of the following:**

   A. Research Notes - short reports dealing with current genetics, breeding and closely related matters that are of possible interest to members.

   B. Germplasm Exchange - a listing of seed stocks that are available or desired. Brief descriptions and gene symbols, if applicable, are useful.

II. **General Guidelines**

   A. Reports should normally not exceed two (2) single-spaced, typewritten or word-processed pages.

   B. Authors are requested to submit electronic copy of their reports by email. The report should be submitted as a word processing file. A follow up email should be sent to see if it was properly received.

   C. Tables and Figures (e.g., *.TIFF, *.PCX, *.GIF, *.JPG, *.WPG) should be included as separate files on the disk even if they are also embedded in the body of the text.

   D. If you are unable to submit your report by email or disk, send a typed copy. CGC will look after re-entering your submission.

III. **Title**

   A. The title should be a precise and concise description of the work.

   B. Avoid the use of meaningless words such as "influence of," "effects of," "results of," "studies on," "evaluation of," "factors involved in," and "tests on."

   C. Begin at left-hand margin. (See Examples I, II and III)

   D. Capitalize first letter of all words except for articles such as "a" and "the," prepositions such as "of," "in," "on," "during," and "between," and conjunctions such as "and" and "with" that are not the first word.

   E. DOUBLE SPACE between Title and By-line.

IV. **By-line**

   A. Author(s) name(s) (first name or initial followed by middle initial and last name). (See Example I)

      1. Names of two or more authors at the same institution are on the same line. (See Example II)

      2. Names of authors in separate institutions are on different lines. (See Example III)

   B. Concise mailing address is on the line below the author(s) name(s). (See Examples I, II and III)

C. TRIPLE SPACE between By-line and Body of Report. (See Example I)

V. **Body of Report** (See Example I)

A. Follow conventional format and include a brief Introduction, essential Materials & Methods, and concise Results and Discussion.

B. DO NOT indent the first word of a paragraph.

C. Use numbers enclosed in parentheses for literature citations.

D. DOUBLE SPACE between paragraphs and between body of report and Literature Cited.

VI. **Taxonomy and Genetic Nomenclature** (See Example I)

A. Taxonomy (See Example I)

1. Give the full scientific names of plants, disease organisms, and insects, along with their authority (and if important, the cultivar name).

2. ***Italicize*** scientific names.

3. Use common names whenever possible.

4. Cultivar names can be preceded by the abbreviation for the word cultivar (e.g., cv. Calypso), or can be set off with single quotes (e.g., 'Calypso').

B. Genetic Nomenclature (See Example I)

1. Names and symbols of genes are subject to the gene nomenclature rules for the Cucurbitaceae. (Robinson et al. 1976. Genes of the Cucurbitaceae. HortScience 11:554-568; CGC Gene List Committee. 1982. Update of cucurbit gene list and nomenclature rules. Cucurbit Genetics Cooperative Report 5:62-66.) These rules were reprinted in the latest CGC Report.

2. Refer to the rules of nomenclature before assigning a name and symbol to a newly described gene in a published report regardless of where it is published.

3. If necessary, consult the CGC Gene List Committee regarding questions of gene names and symbols. Members of the Gene List Committee are listed in the latest CGC Report.

4. ***italicize*** gene names and symbols.

VII. **Literature Cited** (See Example I)

A. List citations in alphabetical order, but numbered consecutively with Arabic numerals followed by a period.

B. Authors are listed after the number; senior author (last name first, by initials), then additional authors (initials first).

C. DO NOT substitute the underline for the author's name when an author is cited more than once, repeat the author's name for each citation.

D. DO NOT indent the second and any subsequent lines of citations, but begin directly below the first letter of the author's last name.

E. DO NOT underline journal titles.

VIII. **Tables** (See Example IV)

A. Tables should document or clarify, but not duplicate, data already given in the text or figures.

B. Large tables can be reduced in size through photoreduction (or reduced font size) in order to fit within the prescribed margins. Photoreductions should be done by the author(s) if possible.

C. Table Anatomy

1. Headnote - contains "Table," then number in Arabic, and a self-explanatory title.

2. Headrule - underscores the headnote; one line.

3. Stubhead - is the head of the first column. Capitalize only the first letter of the first word and any proper nouns.

4. Boxhead - contains the column heads of the rest of the table, and is centered between the stubhead and the right margin. Capitalize only the first letter of the first word and any proper nouns.

5. Boxhead rule - one line under the boxhead to separate it from the main body of the table.

6. Field - is all the information between the boxhead rule and the footrule - - the main body of the table.

7. Footrule - a single underscore to separate the field from the footnotes (if any).

8. Footnotes - are designated with superscript, lowercase letters in reverse alphabetical order (z, y, x, w, etc.), thus avoiding confusion with alphabetical letters used for statistical significance (a, b, A, B).

IX. Figures

A Data presented in tables should not be duplicated in Figures.

B. Figures include graphs and line drawings in black on white paper or on white paper imprinted with light blue lines which will not appear when photographically reproduced, and black and white photographs.

C. Large figures can be reduced in size through photoreduction in order to fit within the prescribed margins. Photoreductions should be done by the author(s) if possible.

D. Captions should be clear, concise and complete.

E. If mailing reports, protect figures with stiff cardboard backing and mark envelope "Do Not Bend."

Examples

Example I

Sources of Resistance to Viruses in Two Accessions of *Cucumis sativus*

R. Provvidenti

Department of Plant Pathology, New York Agricultural Experiment Station, Cornell University, Geneva, NY 14456

Recently we have determined that two accessions of *Cucumis sativus* L. cv. Surinam and cv. TMG-1 are valuable sources of resistance to the most common viruses affecting this species in the U. S.

'Surinam', a cultivar from the South American country of the same name, possesses a single gene (*wmv-1-1*), which confers resistance to watermelon mosaic virus 1 (WMV-1) (2). Following inoculation . . .

(body of report)

...breeders with sources of resistance to four viruses.

Literature Cited

1. Provvidenti, R., D. Gonsalves, and H.S. Humaydan. 1984. Occurrence of zucchini yellow mosaic virus in cucurbits from Connecticut, New York, Florida, and California. Plant Disease 68:443-446.

2. Wang, Y.J., R. Provvidenti, and R.W. Robinson. 1984. Inheritance of resistance to watermelon mosaic virus 1 in cucumber. HortScience 19:587-588.

Example II

Obtention of Embryos and Plants from In Vitro Culture of Unfertilized Ovules of *Cucurbita pepo*

D. Chambonnet and R. Dumas de Vaulx

Institut National de la Recherche Agronomique, 84140 Montfavet, France

Example III

Lack of Resistance to Zucchini Yellow Mosaic Virus in Accessions of *Cucurbita maxima*

R. Provvidenti

Department of Plant Pathology, New York Agricultural Experiment Station, Cornell University, Geneva, NY 14456

R. Alconero

U. S. Department of Agriculture, Agricultural Research Service, Regional Plant Introduction Station, Geneva, NY 14456

Example IV

Table 1. Petiole length (cm) of the first four true leaves of mutant and normal cucumber plants segregating for the short petiole (*sp*) gene.

	Leaf node			
Genotype	**1**	**2**	**3**	**4**
sp sp	1.9	1.8	6.7	3.2
Sp sp	15.0	14.2	16.2	16.1
Sp Sp	15.2	15.9	17.6	17.8

2008 Public Sector Cucumber Research Priority Survey

Yiqun Weng
USDA-ARS, Horticulture Department, University of Wisconsin, Madison, WI 53706

The mission of the Cucumber Improvement Program in the Vegetable Crops Research Unit (VCRU) of USDA-ARS, Madison is to conduct researches to serve the needs of the cucumber industry and consumers. For researchers in a public institution, it is useful to survey their clientele and prioritize their research to address major problems. In December 2008, a national wide survey was conducted to identify priorities for cucumber research in the public sector (see Appendix for the survey design). The questions in the survey were in four categories: diseases, insects, abiotic stresses and other issues. In each category, the respondent was asked to identify and rank in the order of importance of current problems in cucumber production. Write-in space was provided in case the respondents had additional important issues.

The survey was sent to cucumber-related researchers in public institutions (mainly university research and extension faculty), seed companies (cucumber breeders), as well as people working in the cucumber industry. Twenty-one feedbacks were received, of which seven, five, and nine respondents were from the public, private sectors, and the industry, respectively. The results were compiled by inverting the ranks by each respondent where a rank of 1 (top priority) was assigned a value of 5, and a rank of 2 was assigned a value of 4 and so on. Therefore, a surveyed question with the highest value had the highest priority in this category. The results from the public and private sectors, as well as the industry were compiled separately to reflect their different responses to certain questions.

The survey results are summarized in Table 1. The issues in each category were arranged according to the overall ranking of their importance among all respondents. For cucumber diseases, it is clear that downy mildew had the highest priority. Phytophthora fruit rot, angular leaf spot (ALS), cucumber mosaic virus (CMV) and root knot nematode (RKN) are other four with major concerns. Respondents from seed companies also indicated the importance to work with anthracnose and belly rot. Among the major insects, cucumber beetles were ranked the top priority, followed by aphids, pickleworm and thrips. For abiotic stresses, herbicide damage, cold germination, and drought/heat stresses were some important issues. For other cucumber research-related issues (category 4), higher yield was the top priority among public and industry respondents. Improving pre- and post-harvest fruit qualities was also emphasized. Meanwhile, respondents of seed companies ranked 'broadening cucumber genetic diversity' and 'Use molecular markers in marker-assisted selection' as the top priorities. In addition, seed company respondents also emphasized improving fruit nutrition and developing cucumber genomics resources.

In addition to questions asked in the four categories, other issues raised by the respondents during this survey included hybrids for the small cucumber 1A, 1B size market and developing machine harvest system to accommodate this fruit; improve seed vigor; increase fruit per plant; development of parthenocarpic varieties, and finally controlling Length/Diameter ratios with water/fertilizer applications.

To summarize, although rigorous statistical methods were not applied to the survey data, this survey provided very useful information for public sector researchers to prioritize their research to address needs of the cucumber industry in the U. S.

Acknowledgements

The author is grateful to Dr. Mike Havey (USDA-ARS, University of Wisconsin - Madison) and Dr. Gary Taurick (Harris Moran Seed Co) for help in design of this survey. The author also thanks Mr. Brian Bursiek (PPI - Pickle Packers International, Inc) to help distribute the survey among PPI members.

Table 1. Results of Public Sector Research Priority Survey

Categories	Industry		Private		Public		All	
	Weights	Rank	Weights	Rank	Weights	Rank	Weights	Rank
1. Diseases								
Downy mildew (DM)	39	**1**	20	**1**	33	**1**	92	**1**
Phytophthora fruit rot	23	**2**	13	**3**	21	**2**	57	**2**
Angular leaf spot (ALS)	18	**3**	11	**4**	6	**5**	35	**3**
Cucumber mosaic virus (CMV)	9		8	**5**	9	**4**	26	**4**
Root knot nematode (RKN)	6		14	**2**	4		24	**5**
Bacterial wilt (BW)	6		1		11	**3**	18	**6**
Powdery mildew (PM)	9		2		5		16	**7**
Anthracnose	13	**4**	0		2		15	**8**
Belly rot	15	**5**	0		0		15	**9**
Gummy stem blight (GSB)	8		3		2		13	**10**
Watermelon mosaic virus (WMV)	3		4		4		11	**11**
Zucchini yellow mosaic virus (ZYMV)	1		5		5		11	**12**
Cucurbit Yellow Stunting Disorder Virus	2		6		3		11	**13**
Fusarium wilt (FW)	6		3		1		10	**14**
Watermelon strain of papaya ringspot virus	2		0		3		5	**15**
Scab	3		0		0		3	**16**
2. Insect pests								
Cucumber beetles	25	**1**	23	**1**	29	**1**	77	**1**
Aphids	24	**2**	13	**2**	16	**2**	53	**2**
Pickleworm	22	**3**	13	**2**	10	**3**	45	**3**
Thrips	18	**5**	11	**3**	10	**3**	39	**4**
Whiteflies	20	**4**	11	**3**	7	**4**	38	**5**
Spider mites	6		3	**4**	5	**5**	14	**6**
Leaf miners	5		0		5	**5**	10	**7**
Others.	5						5	**8**
3. Abiotic stresses								
Herbicide damage	26	**1**	17	**1**	20	**2**	63	**1**
Drought stress	21	**3**	17	**1**	17	**4**	55	**2**
Cold germination	14	**4**	7	**4**	24	**1**	45	**3**
Heat damage	25	**2**	7	**4**	4	**5**	36	**4**
Chilling damage	7	**5**	13	**2**	14	**3**	34	**5**
Saline stress (salt tolerance)	2		11	**3**	1		14	**6**
4. Other issues								
Higher fruit yield	32	**1**	14	**2**	16	**1**	62	**1**
Improve pre-harvest fruit quality	24	**2**	3	**5**	16	**1**	43	**2**
Broaden cucumber genetic diversity	12	**4**	22	**1**	7	**5**	41	**3**
Use of molecular marker-assisted selection	12	**4**	22	**1**	5		39	**4**
Improve post-harvest fruit quality	15	**3**	0		14	**2**	29	**5**
Improved fruit nutrition	7		11	**3**	7	**5**	25	**6**
Develop cucumber genomic resources	4		9	**4**	11	**3**	24	**7**
Develop GMOs	9	**5**	0		8	**4**	17	**8**

Appendix

2008 Cucumber Research Priority Survey

1. I am a

________	Grower	________	Processor
________	Broker/Marketer	________	Salter
________	Green shipper	________	Public researcher
________	Private researcher	________	Others. Please specify ________

2. My work focuses primarily on

________ Fresh market cucumber ________ Processing cucumber

________ Both

3. If Grower, please check:

________ Less than 100 acres________ 100 to 500 acres________ > 500 acres

4. Areas where you operate (check all that apply):

________ Southeast (AL, AR, FL, GA, LA, KY, MS, NC, SC, TN, VA, WV)

________ Northeast (CT, DE, MA, MD, ME, NH, NJ, NY, PA, RI, VT)

________ Southwest (AZ, NM, OK, TX)

________ Midwest (IA, IL, IN, KS, MI, MN, MO, ND, NE, OH, SD, WI)

________ West (AK, CA, CO, HI, ID, MT, NV, OR, UT, WA, WY)

________ International. Please specify ________________________

For questions 5 to 8, please rank the top 5 topics that you think should be addressed by public-sector research (1 to 5 with 1 = top priority and 5 being lower priority).

5. Diseases

________ Anthracnose (*Colletotrichum orbiculare*)

________ Downy mildew (DM) (*Pseudoperonospora cubensis*)

________ Fusarium wilt (*Fusarium oxysporum f. sp. cucumerinum*)

________ Bacterial wilt (BW) (*Erwinia tracheiphila*)

________ Gummy stem blight (*Didymella bryoniae*, *Phoma cucurbitacearum*)

________ Powdery mildew (PM)(*Podosphaera xanthii*)

________ Belly rot (*Rhizoctonia solani*)

________ Phytophthora fruit rot (*Phytophthora* spp.)

________ Scab (*Cladosporium cucumerinum*)

________ Angular leaf spot (ALS) (*Pseudomonas syringae* pv. *lachrymans*)

________ Root knot nematode (RKN) (*Meloidogyne incognita*; *M. javanica*; *M. arenaria*)

________ Other nematodes. Please specify ________________________

________ Cucumber mosaic virus (CMV)

________ Watermelon mosaic virus (WMV)

________ Zucchini yellow mosaic virus (ZYMV)

________ Watermelon strain of papaya ringspot virus

________ Cucurbit Yellow Stunting Disorder Virus

________ Other diseases. Please specify________________________

6. Insect pests

__________ Cucumber beetles
__________ Spider mites
__________ Pickleworm
__________ Aphids
__________ Whiteflies
__________ Leaf miners
__________ Thrips
__________ Others. Please specify ______

7. Abiotic factors affecting cucumber production

__________ Chilling damage
__________ Drought stress
__________ Saline stress (salt tolerance)
__________ Cold germination
__________ Heat damage
__________ Herbicide damage
__________ Others. Please specify ____________________________________

8. Other issues in cucumber improvement research

__________ Use of molecular marker-assisted selection for cucumber breeding
__________ Broaden cucumber genetic diversity through exploring other *Cucumis* resources
__________ Develop cucumber genomic resources (mapping populations, genetic/physical maps, genome sequencing, double haploid production, ETS ...)
__________ Improve post-harvest fruit quality (brining quality, shelf-life, ...)
__________ Develop GMOs (Genetically Modified Organisms)
__________ Improve pre-harvest fruit quality (shape, color, internal defects, ...)
__________ Improved fruit nutrition (carotenoids content, solid content, nutraceutical ...)
__________ Higher fruit yield
__________ Others. Please specify ____________________________________

9. Other problems not listed. Please specify.

10. Additional comments related to research needs.

A Rapid Spectrophotometric Method to Determine β-Carotene Content in *Cucumis melo* germplasm

Angela R. Davis*, Wayne W. Fish, and Penelope Perkins-Veazie
USDA, ARS, South Central Agricultural Research Lab, Lane, Oklahoma 74555

Abstract: β-carotene is a carotenoid that has antioxidant properties, is a precursor of Vitamin A, and imparts the orange color in some fruits and vegetables. This compound is the major carotenoid in cantaloupe. Because of its health benefits, the β-carotene content in fruits is of interest to the food industry and to melon breeders. Current methods to assay β-carotene content in fruit are time consuming, expensive, and use hazardous organic solvents. In this report, preliminary data is shown for a method to quantify β-carotene content of cantaloupe puree using light absorbance measured with a xenon flash colorimeter/spectrophotometer. Absorbance of twenty seven cantaloupe purees from one variety demonstrated a linear correlation coefficient (R^2=0.7) with β-carotene content determined by hexane extraction/spectrophotometry. This linear correlation shows that this method may be suitable for quantifying β-carotene content in purees of fresh cantaloupe. Since pureeing the sample is the only processing required and no chemicals are needed, the method is rapid, inexpensive and produces no hazardous waste.

Materials and Methods: *Sample Preparation.* All steps were performed under subdued lighting at room temperature. Cantaloupe flesh tissue was cut into approximately 2 to 4 cm cubes. Samples (25 to 500 g) were homogenized in a Waring blender until particle sizes were less than 4 x 4 mm. All samples were then pureed using a Brinkmann Polytron Homogenizer (Brinkmann Instruments, Inc., Westbury, New York) with a 20 mm O.D. blade to produce a uniform slurry with particles smaller than 2 x 2 mm. The samples were not allowed to heat or froth. A water soluble form of β-carotene was diluted in water to use as a control (BASF The Chemical Company, Ludwigshafen, Germany).

Low Volume Hexane Extraction Method (LVH): The low volume hexane extraction method was performed as in Fish et al. (2002). Approximately 0.6 g (determined to the nearest 0.001 g) duplicate samples were weighed from each puree into 2 forty ml amber screw-top vials (Fisher, #03-391-8F) that contained 5 ml of 0.05% (w/v) BHT in acetone, 5 ml of 95% ethanol, and 10 ml of hexane. Purees were stirred on a magnetic stirring plate during sampling. Samples were extracted on an orbital shaker at 180 rpm for 15 min on ice. After shaking, 3 ml of deionized water were added to each vial and the samples were shaken for an additional 5 min on ice. The vials were left at room temperature for 5 min to allow for phase separation. The absorbance of the upper, hexane layer was measured in a 1 cm path length quartz cuvette at 479 nm blanked with hexane. The β-carotene content was then estimated using absorbance at 479 nm and factoring in the sample weight (Zechmeister and Polgar 1943; Beerh and Siddappa 1959; Fish et al. 2002).

Puree Absorbance Method: The puree absorbance method was modified from a lycopene detection method in watermelon and tomato (Davis et al. 2003a, b). Briefly, the Hunter UltraScan XE was standardized as per company specifications each day the instrument was used. Purees were mixed well by gently shaking in a sealed plastic bottle and approximately 20 ml of the sample were immediately poured into a 1 cm, 20 ml SR101A cuvette (Spectrocell, Oreland, PA). Samples were scanned in the transmittance (TTRAN) mode under the following settings: the large reflectance port (1.00"), Illuminant at D65, MI Illuminant Fcw, and observer 10°. The instrument was blanked on the empty cuvette. Triplicate readings were taken. Absorbance at 750 nm was subtracted from absorbance at 520 nm for analysis.

Results and Discussion: *Absorbance of β-carotene standard in water.* A serial dilution in water of a BASF β-carotene standard was performed. An aliquot was read using the LVH method to check for accurate preparation for each dilution. Additionally, each dilution was read on the UltraScan XE and the absorbance was compared to the percent of the standard starting solution and the measured β-carotene concentration using the LVH method. The UltraScan XE readings to the LVH estimated β-carotene concentrations were compared (Figure 1). This figure demonstrates that the BASF standard follows the Beer-Lambert law when diluted in water and when read on the UltraScan XE up to an absorbance of three, which is the ceiling for this instrument. This data also demonstrates that the UltraScan XE provides more consistent readings than the LVH method. This finding indicates that an aqueous fruit puree should also obey the Beer-Lambert law.

Absorbance behavior of puree as related to β-carotene content: Based on spectral results, we investigated the possibility of employing absorbance measurements at 430 nm, 490 nm, and 520 nm for cantaloupe puree as a means to estimate the β-carotene content. Samples included tissue from 27 cantaloupe fruit (1 variety). The absorbance reading of the puree at 520 nm gave a higher correlation with the hexane extraction method.

The absorbance at 520 nm measured for each puree as adjusted for scatter by subtraction of the absorbance at 750 nm was plotted against its β-carotene content as measured by hexane extraction (Figure 2). The scatter adjusted absorbances at 520 nm of the purees appear to obey Beer's law with respect to β-carotene content of the puree. The absorbance reading is linearly correlated with β-carotene content, and the linear least squares fit to the data yields the equation: $y = 23.694x + 5.7785$.

Freeze-thawed samples can not be compared with fresh samples. The freeze-thawed samples exhibit a different conversion equation than fresh samples (Data not shown.). For each level of β-carotene, fresh samples read with a higher absorbance than frozen samples. This is likely due to protein and cell wall breakdown in the frozen tissue

Conclusions: In the search for a rapid and reliable way to quantitate β-carotene levels in cantaloupe tissue for screening large numbers of germplasm samples, we are developing a method that utilizes an instrument that can measure actual light absorbance of compounds in a slurried, aqueous medium. The method is simple, uses no hazardous chemicals, and is faster and less expensive than currently used methods. More cantaloupe varieties are being evaluated to determine the accuracy of this method.

Acknowledgements: We would like to thank Anthony Dillard and Amy Helms for providing valuable technical support.

Correspondence to: Angela R. Davis, United States Department of Agriculture, Agricultural Research Service, South Central Agricultural Research Laboratory, PO Box 159, Lane, OK 74555, USA. Phone: 580-889-7395; Fax: 580-889-5783; email: angela.davis@lane-ag.org

References

Beerh, O.P., Siddappa, G.S. 1959. A rapid spectrophotometric method for the detection and estimation of adulterants in tomato ketchup. *Food Technol.* 13:414-418.

Davis, A. R., Fish, W. W. and Perkins-Veazie, P. 2003a. A rapid spectrophotometric method for analyzing lycopene content in tomato and tomato products. *Postharvest Biol. Technol.* 28:425-430.

Davis, A.R., Fish, W.W., Perkins-Veazie, P. 2003b. A rapid hexane-free method for analyzing lycopene content in watermelon. *J. Food Sci.* 68:328-332.

Fish, W.W., Perkins-Veazie, P., Collins, J.K. 2002. A quantitative assay for lycopene that utilizes reduced volumes of organic solvents. *J. Food Composition and Anal.* 15:309-317.

Zechmeister, L., Polgar, A. 1943. cis-trans Isomerization and spectral characteristics of carotenoids and some related compounds. *J. Am. Chem. Soc.* 65:1522-1528.

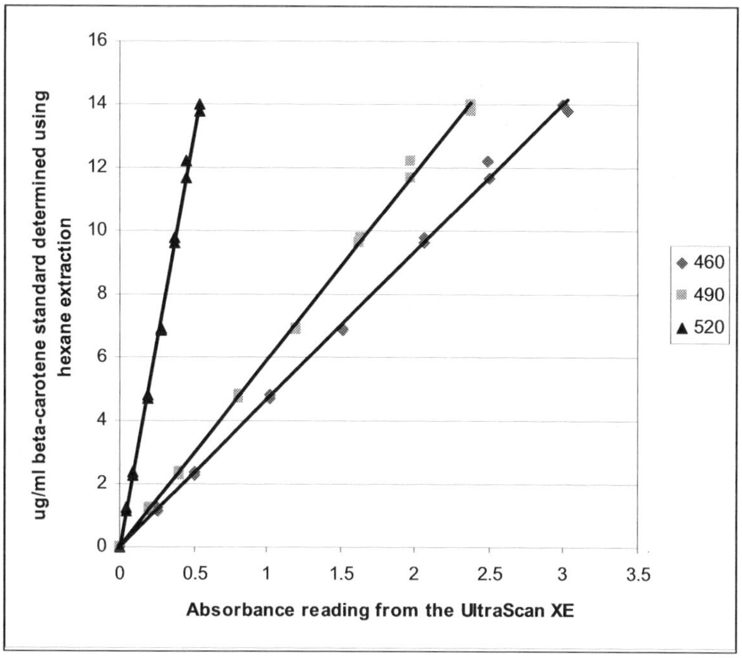

Figure 1: Demonstration of a β-carotene standard obeying Beer-Lambert law when diluted in water and analyzed using the UltraScan XE. Wavelengths 460 and 490 nm were chosen since they provided the highest readings of the scanned standard and the cantaloupe purees. Wavelength 520 nm was chosen because cantaloupe shows a peak at this wavelength. R^2 value for each linear least squared best fit line were all 0.99.

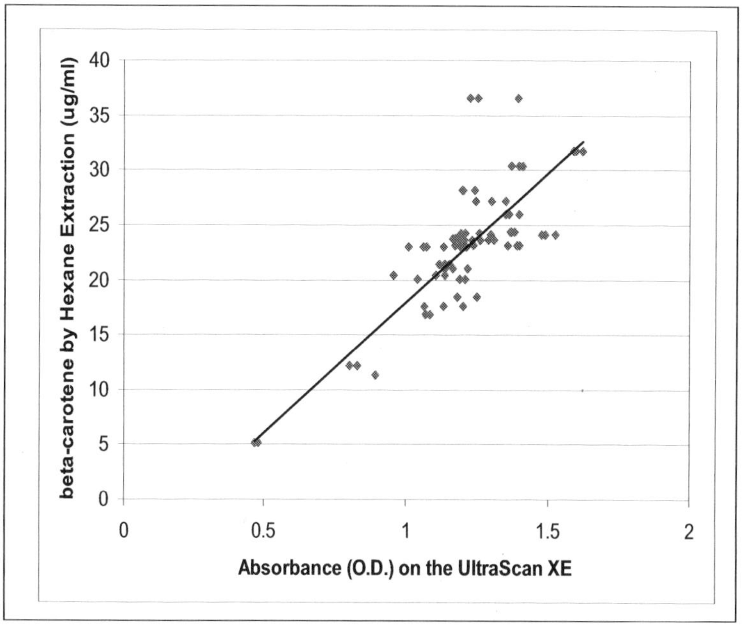

Figure 2: Results of absorbance of 27 cantaloupe purees for β-carotene using the Hunter Lab UltraScan XE. Absorbance is plotted *versus* the β-carotene content of the cantaloupe determined by the low volume hexane method. Absorbance at 520 nm was adjusted for scatter by subtracting the absorbance at 750 nm. The absorbance reading was linearly correlated with β-carotene content, and the linear least squares fit to the data yielded the equation: $y = 23.694x + 5.7785$. The R^2 value for this linear correlation was 0.7.

Green-Fleshed Watermelon Contains Chlorophyll

Angela R. Davis, Penelope Perkins-Veazie
USDA, ARS, South Central Agriculture Research Laboratory, P.O. Box 159, Lane, OK 74555 U.S.A.

Stephen R. King
Vegetable & Fruit Improvement Center, Department of Horticultural Sciences, Texas A & M University, College Station, TX 77843-2133 U.S.A.

Amnon Levi
USDA, ARS, United States Vegetable Laboratory, 2700 Savannah Highway, Charleston, SC, 29414-0000 U.S.A.

Many popular and technical reports on watermelon flesh colors ignore green, an uncommon color. The earliest report of this color mutant that we were able to find dates back more than one hundred years. In this report, inheritance of pink-fleshed vs. green-fleshed genes in watermelon was explored and resulted in fruit intermediate in character, the flesh having a yellowish cast tinged with pink (Card and Adams, 1901). Years later, Whitaker and Davis (1962) listed greenish-white as one of the flesh colors found in watermelon, as well as white, yellow, and red. More recently, Robinson and Decker-Walters (1997, page 84) wrote, "The bland to sweet-tasting flesh is usually red, but may be green, orange, yellow or white in some cultivars or landraces." They also mention, "Citron has white or pale green flesh which is bland or bitter" (p.97). Additionally, green is one of the eight flesh colors and color combinations listed in the Germplasm Resources Information Network for *Citrullus* spp. (USDA, ARS, GRIN). In addition to the rare reports of green color, the compound giving this greenish cast has not been mentioned.

Since some cucurbits have chloroplasts and thus chlorophyll inside their fruit, we surmised that the green may be chlorophyll. We analyzed fifteen *Citrullus* spp. that demonstrated white to greenish-white flesh to determine if chlorophyll is present.

Quantification of chlorophyll was performed on 14 greenish-white watermelon flesh samples. The sample came from citron type PI lines 271769 (8 fruit), 271773 (1 fruit), and 299378 (2 fruit); *lanatus* type PI line 494531 (1 fruit); and volunteers that resembled citron types (3 fruit). Three of the greenish-white flesh samples had no detectable amounts of total chlorophyll and were probably below the detectable limits using our spectrophotometric analysis. The remainder of the samples, including the *lanatus* type, contained very low levels, from 1 to 6 µg/g total chlorophyll. This is roughly 7% of the amount of chlorophyll found in fresh broccoli florets.

Three of the greenish-white samples above (one 271769 and two 271773) were separated using high performance liquid chromatography (HPLC); two white samples were also run using HPLC ('Cream of Saskatchewan' and PI 314655). Comparison of the sample peaks to chlorophyll standards verified chlorophyll peaks in three greenish-white samples. The white-fleshed watermelon samples gave no detectable chlorophyll peaks. These results suggest that chlorophyll is imparting the green tint to greenish-white-fleshed *Citrullus* spp.

We know that carotenoids in watermelon are packaged in chromoplasts, similar to red tomato (Fish, 2006; Harris and Spurr, 1969), but we found no reports that watermelon flesh contains chloroplasts. Our finding suggests that *Citrullus* spp. containing chlorophyll also contain chloroplasts.

If there are indeed chloroplasts in watermelon flesh, this generates more questions. Is chlorophyll present in all watermelon or only the ones with a greenish tint? How much chlorophyll does watermelon have the potential to make? Is chlorophyll or the other components of chloroplasts, such as violaxanthin, affecting the complexity of color in red, yellow, and orange watermelon? Are the chloroplasts functional? Does light penetrate the rind and does the presence of chlorophyll in the flesh help the fruit synthesize energy and sugars?

Materials and Methods: *Plant material.* Ripe watermelons of white to greenish-white were grown at Lane, OK, and in College Station, TX from 2005-2007. Five Plant Introduction lines (PI) (271773, 271769, 299378, 314655, and 494531), one open pollinated variety ('Cream of Saskatchewan'), and three volunteers of unknown origin were evaluated for chlorophyll. Fruits were cut the day of harvest and flesh tissue was extracted from the heart.

Sample Preparation: Immediately after collecting heart tissue the samples were frozen and stored at -80°C until processed for spectrophotometric and HPLC analyses. Tissue (~30 g) was homogenized using a Brinkmann Polytron Homogenizer (Brinkmann Instruments, Inc., Westbury, NY) with a 20 mm O.D. blade to produce a uniform slurry with particles smaller then 3 mm^3. For HPLC analyses, samples were concentrated by centrifugation and then extracted. Samples analyzed using a spectrophotometer were not concentrated before extraction. Samples were extracted using a modified acetone extraction method (Lichtenthaler, 2001).

HPLP and spectrophotometric analysis: A subset (three) of the samples tested above and two white-fleshed samples were concentrated and then extracted as above and analyzed following HPLC methods previously described (Craft, 2001). Samples were filtered using 0.45 µm PTFE syringe filters (Daigger, Vernon Hills, IL) into 2-ml amber crimp-top vials (Daigger, Vernon Hills, IL), then loaded onto an Agilent model 1100 high performance liquid chromatography system equipped with autosampler, photodiode array detector, and integration software (Agilent model 1100,Wilmington, DE). A C30 YMC carotenoid column (4.6 x 250 mm) and YMC carotenoid guard column S-3 (4.0 x 20 mm) (Waters, Milford, MA) was used. A gradient method with three solvent mixtures was used for separation. Solvent mixtures of (A) 90% methanol, 10% ddi water containing 0.5% triethylamine and 150 mM ammonium acetate, (B) 99.5% 2-propanol, 0.5% triethylamine, and (C) 99.95% tetrahydrofuran, 0.05% triethylamine were applied as follows: initial conditions 90% solvent A plus 10% solvent B; 24 minute gradient switched to 54% solvent A, 35% solvent B and 11% solvent C; final gradient conditions were 11 minute gradient of 30% solvent A, 35% solvent B, 35% solvent C, held for 8 minutes. The mobile phases were returned to initial conditions for 15 min. Injection volumes of 100 µl were used for samples and standards. Chlorophyll A and B standards were obtained from Sigma (St. Louis, MO) and were used for peak verification.

Chlorophyll A and B were quantified using the spectrophotometric assay in Wrolstad et. al. (2004).

Literature Cited

Card, F.W. and G.E. Adams. Rhode Island Sta. Rpt. 1902, pp. 227-244, pls.9. As reported in: Experiment Station Record. 13:1901-1902. Allen, E.W., et. al. (eds.). Washington Government Printing Office. p. 741.

Craft, N. 2001. Chromatographic techniques for carotenoid separation. Current Protocols in Food Analytical Chemistry. F2.3.1-F2.3.15.

Fish, W.W. 2006. Interaction of sodium dodecyl sulfate with watermelon chromoplasts and examination of the organization of lycopene within the chromoplasts. J. Ag. Food Chem. 54:8294-8300.

Harris, W.M. and A.R. Spurr. 1969. Chromoplasts of tomato fruits. II. The red tomato. Am. J. Bot. 56:380-389.

Lichtenthaler, H.K. and Buschmann, C. 2001. Extraction of photosynthetic tissues: chlorophylls and carotenoids. Current Protocols in Food Analytical Chemistry. F4.2.1-F4.2.6.

Whitaker, T.W. and Davis, G.N. 1962. Taxonomy and cultivars. In: Cucurbits, botany, cultivation, and utilization. P. 38. Whitaker, T.W. and Davis, G.N. (eds.) Interscience Publishers, Inc. New York, NY.

Robinson, R.W. and D.S. Decker-Walters. 1997. Major and Minor crops. In: Cucurbits. Pp. 84 & 97. R.W. Robinson and D.S. Decker-Walters (eds.) CAB International, Wallingford, UK.

USDA, ARS, National Genetic Resources Program. Germplasm Resources Information Network - (GRIN). [Online Database] National Germplasm Resources Laboratory, Beltsville, Maryland. Available: http://www.ars-grin.gov/cgi-bin/npgs/html/codes.pl?151007 (11 Sept. 2008)

Wrolstad R.E., T.E. Acree, E.A. Decker, M.H. Penner, D.S. Reid, S.J. Schwartz, C.F. Shoemaker, D.M. Smith, and P. Sporns. 2004. Current Protocols in Food Analytical Chemistry. John Wiley and Sons, New York, NY.

Acknowledgments: We would like to thank Amy Helms, Julie Collins, and Sheila Magby for providing valuable technical support.

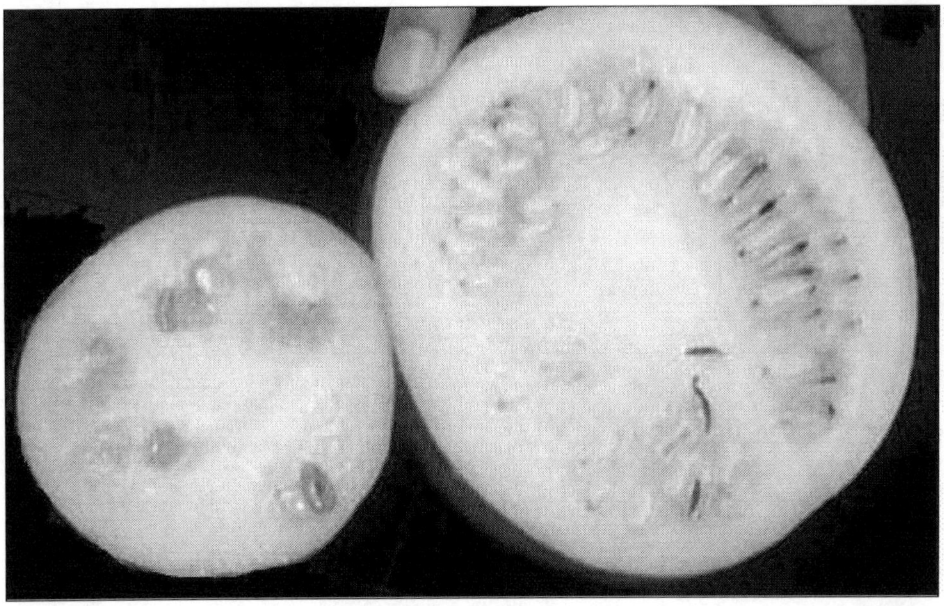

Figure 1. Green-fleshed watermelon on the left and white-fleshed watermelon on the right. The white-fleshed watermelon also demonstrates the Egusi seed characteristic.

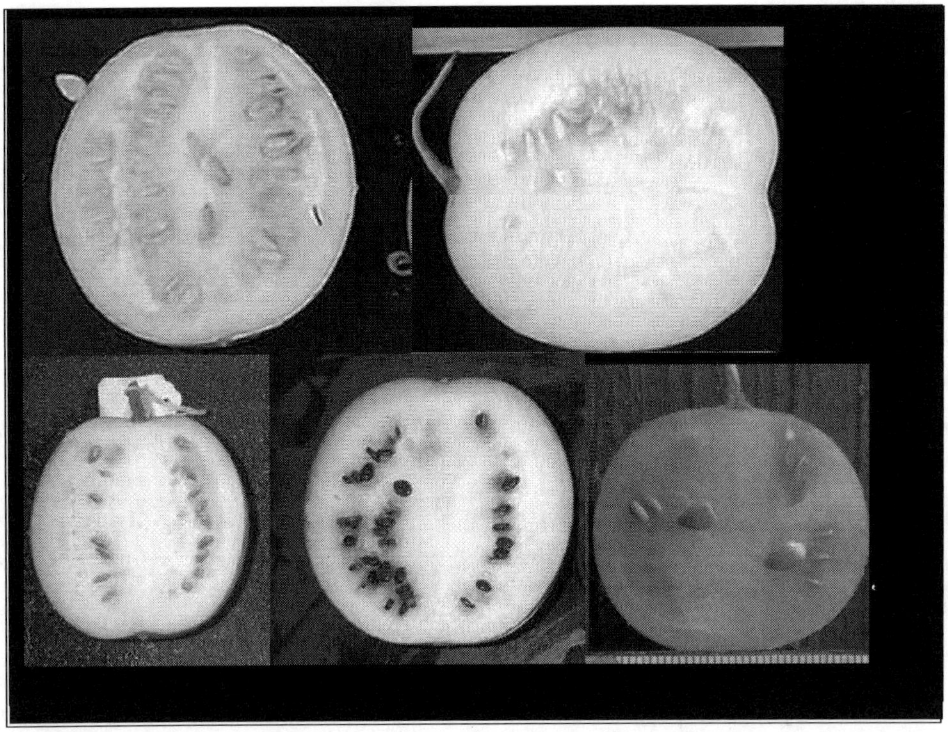

Figure 2. A range of watermelon fruit colors staring in the upper left-hand corner with pure white flesh, and ending in the lower right-hand corner with intense green. Intermediate amounts of green color as shown in the remaining three fruit.

A New Dwarf Mutant dw-4 in Watermelon

Hua YANG, Yong-gang LI, Ding-xin YANG & Jie YANG
Private Research, A8-1Xilaiyuan, Jinyu Road, Haikou, Hainan 570203, P. R. China

Abstract: A spontaneous dwarf mutant was discovered from inbred watermelon line '5-6y'. Genetic study showed that the dwarf mutant is genetically stable and the dwarf trait is inherited as a single recessive gene. Allelism test showed that the new dwarf gene is not allelic to the two known dwarf genes dw-1 and dw-2. The allelism test of this new mutant with dw-3 was not possible because the genetic stock of dw-3 is no longer available. The phenotype of this mutant appears different from dw-3 and we suggest that the new mutant gene is named as dw-4.

Key word: Watermelon; Dwarf, Gene

Introduction

A short internode plant was observed in a watermelon line '5-6y' grown in a watermelon winter nursery in Hainan Island in the winter of 2004. The short internode was stable in the self-pollinated progeny of the mutant plant. The line with short internode was then named as 'd5-6y'. The wild type line '5-6y' has a main stem length of 240 cm, and internode length of 8 - 12 cm. There is potential of branching at every node of the vine in the wild type. However, the mutant line 'd5-6y' has main stem length of 80 - 90 cm, internode length of 4 - 5 cm, and the plant has fewer branches, with 4 - 6 branches per plant. Leaf and fruit of the mutant plant are smaller than those of the wild type. The mutant plants produce normal flowers. Both the mutant and wild type have delayed green cotyledon and growing points (Figure 1 and 2).

Materials and Methods

Experiment 1: Crosses were made to determine the genetics of the dwarf mutant in 'd5-6y'. The F_1's are P1 x P2, P1 x P3, and P1 x P4. BC1's: (P1 x P2) x P1, (P1 x P3) x P1, and (P1 x P4) x P1 where P1 = 'd5-6' (short vine mutant), P2 = PL (normal vine wild type), P3 = Sugarlee (normal vine wild type), and P4 = Allsweet (normal vine wild type). A χ^2 test was used to test the goodness of fit.

Experiment 2: Crosses were made to test the allelism between the new dwarf mutant and reported dwarf mutant dw-1 and dw-2 (Guner, N. and T. C. Wehner. 2004). The reciprocal F_1's were P1 x P2, P2 x P1, P1 x P3 and P3 x P1, where P1 = 'd5-6y' (short vine mutant), P2 = Bush Sugarbaby (dw-1/dw-1, Figure 3), and P3 = BR15d (dw-2/dw-2, Figure 4). A χ^2 test was used to test the goodness of fit.

All the populations were phenotyped on plants grown in the open field.

Results and Discussion

As shown in Table 1, the F_2 progenies segregated in a 3:1 ratio (normal to dwarf), and the BC_1 progenies segregated 1:1. These genetic test results show that the dwarf trait in mutant line 'd5-6y' is conferred by a single recessive gene.

All the reciprocal F_1's made between the new mutant and the genetic stocks of dw-1 and dw-2 had the normal plant type (data not show). This means that the new dwarf mutant gene is neither allelic to dw-1 nor dw-2.

A dw-3 mutant was reported in watermelon (Huan, 1995). However, our effort to get the genetic stock of dw-3 mutant was not successful. The genetic stock of dw-3 is no longer available according to Huan. Therefore the allelism test between the new dwarf mutant and dw-3 is impossible.

The dw-3 mutant (Figure 5) was derived from watermelon line DMSW. The dw-3 mutant plant also shows male-sterility and non-lobe leaf. The male-sterility in dw-3 is conferred by the gene ms^{dw}.

Due the dramatic phenotypic different between the dw-3 mutant and the new dwarf mutant 'd5-6y' and the inability of testing allelism between dw-3 and the new mutant, we propose a new gene dw-4 for the new mutant.

References

Guner, N. and T. C. Wehner. 2004. The genes of watermelon. HortScience 39: 1175-1182

Huan, H. X., 1995. Study of Dwarf Male-sterile Watermelon. China Watermelon & Melon 3: 6-9 (in Chinese)

Table 1. Segregation of F_2 and BC_1 Progenies Derived from the Mutant Line 'd5-6y' and the Wild Type Lines

Cross	Generation	Total Plants	Dwarf Plants	Normal Plants	Dwarf/Normal	X^2-value
d5-6y X PL-1	F_2	516	137	379	1:2.77	0.6615
d5-6y X PL-2	F_2	509	127	382	1:3.01	0.0007
d5-6y X PL-3	F_2	610	155	455	1:2.94	0.0546
d5-6y X PL-4	F_2	172	41	131	1:3.20	0.1240
d5-6y X PL-5	F_2	277	77	200	1:2.60	1.1564
PL X d5-6y-1	F_2	233	54	179	1:3.31	0.4134
PL X d5-6y-2	F_2	253	62	191	1:3.08	0.0329
d5-6y X Sugarlee-1	F_2	276	67	209	1:3.12	0.0773
d5-6y X Sugarlee-2	F_2	366	83	283	1:3.41	1.0528
d5-6y X Sugarlee-3	F_2	490	111	379	1:3.41	1.4395
Sugarlee X 5-6y-1	F_2	425	89	336	1:3.78	3.7341
d5-6y X Allsweet	F_2	346	93	253	1:2.72	0.6513
Allsweet X d5-6y	F_2	539	146	393	1:2.69	1.2523
(d5-6y X PL) X d5-6y	BC_1	322	177	155	1:0.88	1.8137
(d5-6y X PL) X d5-6y	BC_1	250	121	129	1:1.07	0.2560
(d5-6y X PL) X d5-6y	BC_1	300	157	143	1:0.91	0.6533
(d5-6y X PL) X d5-6y	BC_1	162	73	89	1:1.22	1.5802

$X^2_{0.01(df=1)}=6.635$ $X^2_{0.05(df=1)}=3.841$

Figure 1. Plant phenotype of the new mutant (dw-4, top plant) and the wild type (bottom plant).

Figure 2. Plant morphology of the new dwarf (dw-4) mutant plant grown in open field.

Figure 3. Plant morphology of dw-1 mutant, CV Bush Sugarbaby

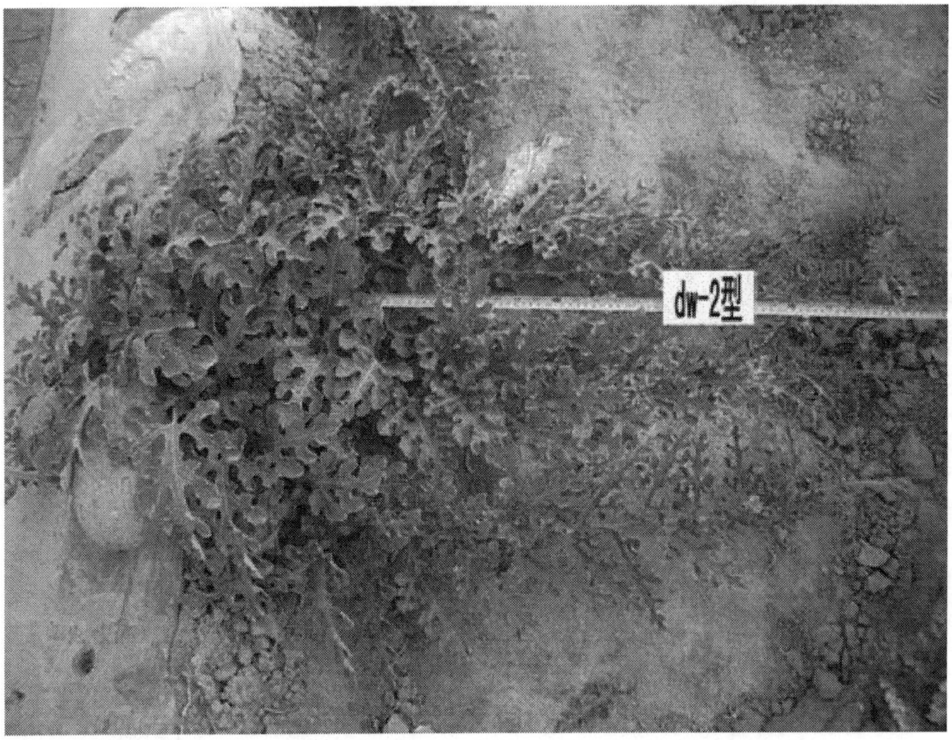

Figure 4. Plant morphology of dw-2 mutant plant, CV BR15d.

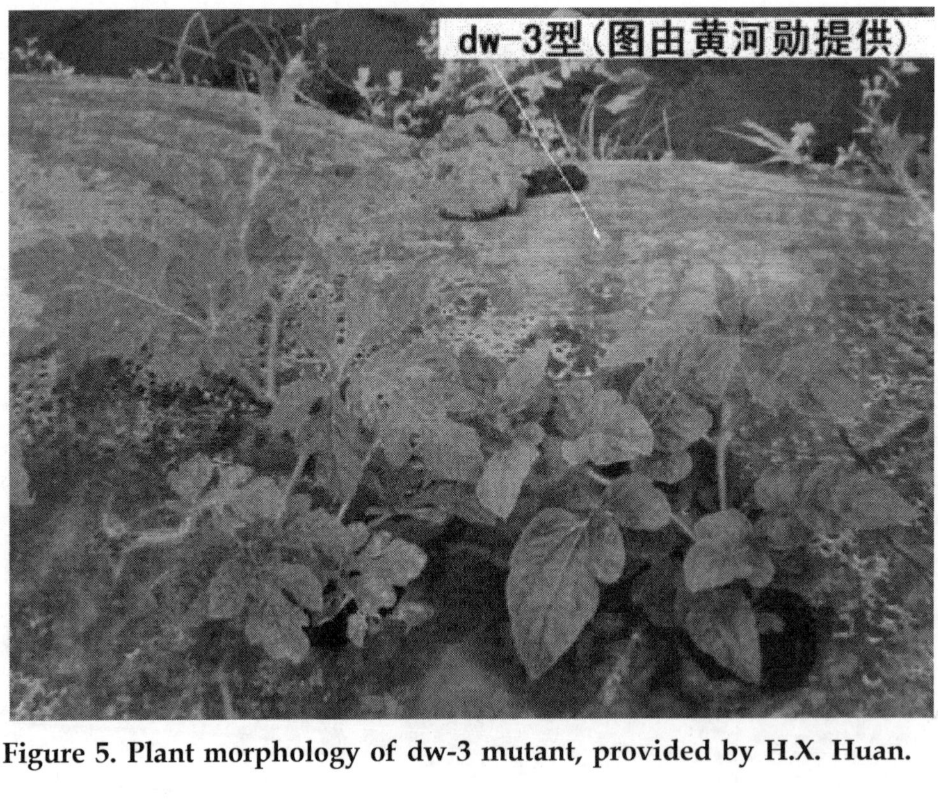

Figure 5. Plant morphology of dw-3 mutant, provided by H.X. Huan.

A Apetalous Gynoecious Mutant in Watermelon

Yong-gang LI, Hua YANG, Jie YANG, Ding-xin YANG
Private Research, A8-1Xilaiyuan, Jinyu Road, Haikou, Hainan 570203, P. R. China

A gynoecious (gy) mutant was previously described in watermelon (Jiang and Lin, 2007; Fig. 1). This original gy mutant will eventually produce some perfect flowers which will allow the genotype to be selfed and maintained.

A apetalous gynoecious mutant was observed from gynoecious line 'Mi Guo' grown in the winter nurseries in Sanya, Hainan in 2006. This mutant was cross-pollinated with wild type genetic stock and the F1 progeny had normal flowers and petals. Apetalous gynoecious individuals were observed in the F2 progeny with the phenotype of original mutant (Fig. 2). Limited data suggest that the apetalous trait is conferred by a single recessive gene. Because of the strong gynoecity, we are not able to induce male flowers from the mutant to self-pollinate the mutant. The mutant is currently maintained by crossing the apetalous gynoecious plant with monoecious wild type or androgynous plant of a breeding line. The apetalous gynoecious mutant has normal female fertility and produces seed normally.

More genetic studies are being conducted to understand the genetics of the apetalous trait and its relationship with gy trait.

Acknowledgement: Authors wish to thank Dr. Xingping Zhang of Syngenta Seeds, Inc. for his encouragement and help in preparation of this article.

Reference

Jiang, Xiantao and Depei Lin, 2007, Discovery of Watermelon Gynoecious Gene gy, Acta Horticulturae Sinica, 34(1): 141 - 142

Figure 1. Original gynoecious watermelon mutant previously described.

Figure 2. Flower characteristic of the apetalous gynoecious mutant. Picture was taken from field grown plants.

Breeding Classic *Cucurbita maxima* Buttercup Squash for Increased Genetic Diversity

Nicki Dallmann and Michael Dallmann
The Evergreen State College Organic Farm, 2712 Lewis Rd. NW, Olympia, WA 98502

This article was received for the Cucurbit Genetics Cooperative Report No. 30 but is printed here in the No. 31-32 issue.

Introduction: The unique ability for cucurbits, naturally outcrossers, to withstand inbreeding has allowed pedigree line breeding techniques to become the industry standard for development of new squash varieties, buttercup included (1; 7). Self-fertilization is preferred because it ensures rapid homozygosity and the fixation of desirable characteristics within the variety (1). A method of recurrent selection breeding was developed with the goal of creating a new variety of *C. maxima* buttercup squash that competed with or outperformed other market varieties, yet maintained maximum heterozygosity. This would be achieved by avoiding self-pollination at critical stages of the breeding project and maintaining adequate numbers of progeny over the selection process.

Materials & Methods: The seed from the original population was developed in 1994-96 by John Navazio and John Schneeberger at Garden City Seeds in Hamilton, Montana (5). The project sought to combine the characteristics of cold soil vigor, classic buttercup shape, early fruit maturity, and high fruit quality by combining a small-sized, well-maintained, market version of the standard variety Buttercup (Cooper's Seeds, Auckland, New Zealand) with the vigorous, early maturing farmer variety Selma's Buttercup, bred by Selma Gunderson of Buffalo, South Dakota. The latter was segregating for shape, epidermal color, and fruit quality. A minimum of a dozen plants of each of the two varieties were allowed to open-pollinate and seed was bulked. This seed was planted into cool soil at least two weeks before commercial crops were planted in the Bitterroot Valley and seedlings were selected for early, strong emergence and vigor. The resultant population of 5,400 to 6,000 plants was allowed to open-pollinate. At early harvest (Sept 1-10) plants with at least three fruit at full maturity with were marked for harvest. At frost at least 350 fruit were harvested from at least 250 selected plants. Seed was harvested after four weeks of curing in the fruit and was bulked. This process was repeated for a second season and the population named Montana Maxima Early Composite (MMEC). Starting with this population selection toward a new variety was started in Olympia Washington ten years later with a progeny test. Refinement of a promising family over a three-year period led to a possible variety that exhibited a high percent of phenotypic uniformity for classic buttercup characteristics and high quality fruit, and also demonstrated a higher potential for heterozygosity.

Field plots at The Evergreen State College Organic Farm were arranged in 6′ x 6′ grids with 6′ between mound centers. The placement of squash full-sib families and dimensions of the plot varied from year to year with rotation of the field and available space. Three feet wide pieces of black landscape cloth covered in-row spacing for weed suppression. Between-row spacing was rototilled as needed. Stakes were placed in the mounds to indicate plant families. Full-sib families were direct seeded in early June and harvested in mid October. Fruit harvesting, curing and storage were performed according to Coleman (3). Seed processing and drying followed the methods outlined by Ashworth (2). Once seeds were completely dried, they were placed into labeled envelopes.

To achieve healthy population size 50 plants were grown in 2005, ~300 in 2006 and ~250 in 2007 (6). All pollinations were made by hand to avoid contamination from unwanted specimens and followed the procedure outlined by Ashworth (2). In 2005 male and female flowers were randomly crossed among different plants as they became mature, a method used to reinvigorate the seed through cross-pollination. Included in the field in 2005 were 39 MMEC seeds and 15 market class C. maxima varieties. The family referred to as #8 was originally a cross between 'Bonbon' (F1) (Johnny's Selected Seeds, Winslow, Maine) and a select MMEC plant. In 2006 twenty-one seeds were planted from each of the 12 full-sib families from this cross in 2005. A progeny test was performed by making random pollinations between the 21 siblings within a full-sib family. The best fruit from the four families (#107, #307, #707, and #807) that exhibited the highest performance based on cumulative characteristics were saved for seed.

There were eight superior plants from family #806. In 2007, nine seeds from each hand-pollinated fruit from these eight superior #806 plants were sown and plants within the same family were crossed amongst each other randomly, permitting possible crossings between cousins, siblings, and self-pollinations. Seed from nineteen hand-pollinations were harvested from #8 for future breeding purposes. A pedigree was kept of all male and female parents, their respective performance in the field, and progeny. Squash characteristics were recorded across a number of categories for 2005, 2006 and 2007 including days to seedling emergence, germination rate, days to first pistillate flower, vigor, yield, and fruit quality. Fruit quality was tested in 2007 with the use of taste testing and Brix readings with a refractometer (4).

Results & Discussion: Only family #8 showed a promising consecutive increase in uniformity required for the development of a new variety, with seeds saved from eight out of 23 fruits (~30% uniformity) in 2006 and 19 out of 23 fruits (~60% uniformity) in 2007. Brix readings taken for the four families grown in 2007 were by the far the strongest determining factor for the continued breeding of family #8. The average °Bx RDS for #8 was 12% with the 19 fruit saved exhibiting sweet to super-sweet flavors and an ideal texture characterized as chalky/smooth. When tested, the market variety 'Bonbon', grown as a control, averaged 13% °Bx RDS. The highest Brix reading for 'Bonbon' fruit was 17°Bx RDS and the lowest was 8°Bx RDS. The highest Brix reading for #807 was 16°Bx RDS and the lowest was 6.5°Bx RDS. The highest average percentage reached by other composite families was 10%, and all exhibited low uniformity for desirable texture and sweetness.

Of the four families tested in 2007 for field performance, #8 ranked average to low across several categories. Further field analysis can be made in the future through vigorous comparisons between the composite cross and current market varieties. Family #8 germinated eight days after direct seeding and had an 88% germination rate. It had the second highest germination rate, yet was the slowest in average days to seedling emergence, trailing the others by a half day at nine-and-a-half days. For average days to first pistillate flower #8 took the longest at 57 days, with the shortest of the other three families achieving their first female flowers after 55 days. There is a noticeable difference in average days to first pistillate flower and seedling emergence between 2006 and 2007, with numbers increasing for all MMEC derived families in 2007. Average days to seedling emergence for family #8 increased by one-and-a-half days from 2006 to 2007, and average days to first pistillate flower increased from 55 days in 2006 to 57 days in 2007. This is likely due to a change in weather conditions and soil patterns. Family #8 ranked the second highest in yield with an average of one-and-a-quarter fruits per plant and had an average weight of 1.25 lbs a fruit, the lowest of other MMEC derivatives. The target weight for a new variety of buttercups squash is 2-4 lbs. Field results for #8 indicate the need for careful monitoring of field performance in the future combined with further selection for vigor, early seedling emergence and pistillate flower emergence.

Breeding work to further refine #807 to be released as a commercial variety will be conducted through participatory plant breeding research at the Organic Seed Alliance (OSA), Port Townsend, Washington with the cooperation of Washington farmer-breeders and Prescott College, Prescott, Arizona. Germplasm release and distibution will be handled by the Organic Seed Growers and Trade Association (information is available at the OSA website <seedalliance.org <http://seedalliance.org/>>).

Literature Cited

1. Allard, R. W. 1960. Principles of plant breeding. John Wiley & Sons, Inc., New York.
2. Ashworth, S. 2002. Seed to seed: seed saving and growing techniques for vegetable gardeners. 2nd ed. Seed Savers Exchange, Decorah, IA.
3. Coleman, E. 1999. Four-season harvest: organic vegetables from your home garden all year long. Chelsea Green Publishing Co., White River Junction, VT.
4. Harrill, R. 1998. Using a refractometer to test the quality of fruits and vegetables. Pineknoll Publishing, Keedysville, MD.
5. Navazio, J. 1995. Varietal improvement of winter squash through recurrent selection (USDA Small Business Innovation Research grant proposal). Garden City Seeds, Hamilton, MT. Research Department.
6. Navazio, J. and Morton, F. 2003. Workshop notebook for: organic vegetable seed production for the farm and garden. Olympia, WA: The Evergreen State College, Organic Farm.
7. Whitaker, T. W. & Robinson, R. W. 1986. Squash breeding. In M. J. Bassett (Ed.), breeding vegetable crops (p.210- 238). AVI Publishing Company, Inc., Westport, CT.

Performance of Zucchini Yellow Mosaic Virus Resistant 'Golden Delicious' Type Pumpkin Hybrids

James R. Myers and Deborah Kean
Department of Horticulture, ALS 4017, Oregon State University, Corvallis, OR 97331 USA

Rebecca Brown
Department of Plant Sciences, University of Rhode Island, Kingston, RI 02881 USA

The *Cucurbita maxima* 'Golden Delicious' (GD) type is the preferred pumpkin grown for processing and culinary seed production in the Willamette Valley of Oregon. Processors value their large size (five - 10 kg), thin, red-orange skin (which will not blemish processed product) and high quality flesh (7% soluble solids and 10% total solids). Culinary seed producers prefer GD for its large, plump and attractive white seeds. Over half the GD pumpkin production in the Willamette Valley is for culinary seed, which is most often exported to Pacific Rim markets.

Epidemics of zucchini yellow mosaic virus (ZYMV) occur in the valley every few years. Epidemiology of the virus in the Pacific Northwest is poorly understood. Alternate hosts have not been identified, especially ones that would allow the virus to overwinter from year to year. In an outbreak year, the virus is found first along the Columbia River, and then moves southwards into the Willamette Valley (3). ZYMV epidemics are a problem for pumpkin growers and processors because virus infection of the fruit reduces quality and can cause rejection of fields with too many virus-infected fruit. Fruit symptoms include reduced size, abnormal shape and green patches of skin that are visible on mature fruit. The latter symptom is particularly problematic to processors because the flesh beneath green areas will not ripen, thus not meeting processor specifications for soluble and total solids, and the green skin is a visible contaminant in the processed product.

Immature ovaries and developing fruits of GD are yellow as is typical of pumpkins possessing the *B* gene (4), but fruits ripen to a red-orange color probably conditioned by *Rd* (2). The B^{max} of *C. maxima* is at a different locus from *C. moschata* and *C. pepo* (6), and has been described as completely dominant in contrast to the bicolor pattern of heterozygotes of *C. moschata* and *C. pepo* (5). At maturity, GD fruit exhibit a faint green ring surrounding the stylar scar at the blossom end. When fruit becomes virus infected, the green pigment suppressing effect of *B* and *Rd* is reversed, allowing the expression of a larger green patch at the blossom end, and streaks and patches elsewhere on the fruit (Figure 1).

In 1998, we initiated a program to introgress ZYMV resistance from *C. equadorensis* into *C. maxima* (1). Virus resistance is quantitative, but appears to be controlled by major gene(s). From 1998 to 2001, a cyclical scheme of backcrossing resistant lines to GD followed by intercrossing was conducted for five generations (1). Six generations of selfing followed, resulting in 45 virus resistant inbred lines. Most lines had orange fruit (ranging from yellow-orange to pink, to red-orange), but a few were dark green. The orange-fruited types typically had small fruit size whereas the green skinned types had the more desirable large fruit size. Because inbreds were intended for use in an F1 hybrid production program, we established a yield trial to evaluate hybrid horticultural potential. We tested two sets of materials: crosses of resistant inbreds to GD and crosses between green- and red-orange fruited, virus resistant inbreds. One hypothesis was that virus resistance would be intermediate in crosses between resistant inbreds and susceptible GD. A second hypothesis was that crosses between large green-fruited virus-resistant inbreds and small orange-fruited virus-resistant inbreds would produce medium to large orange-fruited virus-resistant hybrids. We report here the results of a field trial where these hypotheses were tested.

Materials and Methods

Plant materials: Open-pollinated 'Golden Delicious', maintained in our breeding program, was used as the check. Twenty-one virus-resistant inbreds (Table 1) were crossed to GD and to each other to produce 24 unique cross combinations. GD was crossed to a selection of orange-fruited inbreds (Table 2) without regard to which line was the maternal parent. In a second set of materials, one of three X1- inbreds with green skin color were crossed to orange-fruited inbreds. Like the GD crosses, the X1- inbreds were used as the female in some and the male in other crosses.

Trial conditions: Plants were started from seed in the greenhouse on May 13, 2008 in 7.6 cm (3 in) pots using SB40 (Sun Gro Horticulture, Bellview, WA) potting mix supplemented with Apex 14-14-14 slow release fertilizer. Plants were transplanted to the field (Chehalis silt loam) at the Lewis Brown Research Farm, Corvallis, OR on June 6 with five plants per plot. Space between rows was 3.38 m (11 ft) with 92 cm (3 ft) between plants. Plots were arranged in a randomized complete block with three replications. Transplants received 450 lb a^{-1} (504kg ha^{-1}) 12-29-10-4 (N-P-K-S) fertilizer banded into the row just prior to planting. Transplants were irrigated immediately after planting, and the trial received weekly irrigation of approximately 25 mm. Plots were harvested on Oct. 22 and fruit were counted and weighed on an individual plot basis.

Virus inoculation: The ZYMV isolate was originally obtained from Phil Hamm, Hermiston Research and Extension Center, Hermiston, OR. It was stored in frozen (-80C) tissue of 'Honey Boat' (*C. pepo*) Delicata winter squash until use. Virus inoculum was prepared by grinding approximately 10 g of frozen tissue in 100 ml potassium phosphate buffer (2.6 mM monobasic potassium phosphate, 0.047 mM dibasic potassium phosphate, pH 8.5) with 250 mg carborundum powder with a mortar and pestle for one minute. Two-week old 'Honey Boat' plant primary leaves were rub-inoculated using the pestle dipped in inoculum solution. Plants were grown for one month and monitored for symptom expression prior to being used for field inoculation.

Susceptible spreader rows of 'Honey Boat' were direct seeded at the time of transplanting of the GD trial. Spreader rows were planted on the outside of the yield trial and every two rows within the trial. Inoculum for the spreader rows was prepared from greenhouse-infected plants. A Waring blender was loosely packed with symptomatic leaves and about 750 ml of phosphate buffer stored on ice was added and the mixture was blended on the high setting for three minutes. The solution was filtered through three layers of cheesecloth, and was then decanted into an electric paint sprayer modified for large scale virus inoculation. Plants in the spreader rows were inoculated with the paint sprayer when they had at least one expanded primary leaf. Inoculation was considered effective when the paint sprayer left a water-soaked area on the inoculated leaf. We relied on natural aphid transmission to move the virus from the spreader rows into the yield trial. At the time of our first reading on July 10, ½ to ¾ of the GD plants were infected, and by one month later, all GD plants showed virus symptoms (data not shown).

Statistical analysis: Data were analyzed using PROC GLM of SAS (Cary, NC) and means were separated using Fisher's protected least significant difference (LSD). To determine whether differences in virus infection was observed when hybrids had one vs. two parents contributing resistance, LS means were calculated, and the null hypothesis that all means were equal was tested.

Results and Discussion

Yields of the hybrids were generally high, with net yield ranging up to 51 MT ha^{-1} (Table 2). Trials from additional environments would be needed to validate these yields. Fruit weight was generally satisfactory with most hybrids achieving an average fruit weight of 4.5 kg, the minimum sought by processors. Generally, the GD x inbred crosses produced smaller fruit than the inbred x inbred crosses. The GD check was heavily infected with virus, which greatly reduced marketable fruit number and weight (Table 2). Most fruit from this cultivar exhibited typical symptoms of the virus infection, including green patches, misshapen and warty fruit (Figure 1). Symptoms were less severe to nonexistent in the experimental hybrids. Experimental hybrids generally had significantly higher marketable yields under disease pressure compared to GD, however, some experimental lines did have up to 50% of total fruit weight in culls. Culls were considered to be immature fruit, and fruit with a high percentage of the skin with green color. The green fruit color was the result of either virus infection (predominantly in GD and GD crosses), and/or by incomplete dominance of the genes controlling fruit color in green x orange skinned crosses (Figure 2). Partially green fruit color was also observed in green x orange hybrids grown in the absence of the virus (data not shown). *Rd* is epistatic and partially dominant to other fruit colors (2, 4) and is probably the gene responsible for the large green blossom ends observed in the green x orange crosses. We did not expect GD x orange hybrids to show any green at the blossom end since it was thought that both parents were homozygous for *B* and *Rd*. One possibility is that *Rd* had been lost from some inbreds, but the red-orange skin color of all inbreds used in this study does not support this idea. GD crosses had 10 - 28% (mean = 21%) cull fruit whereas green x orange crosses ranged from 14 - 49% (mean = 33%). We attribute the higher cull frequency in hybrids that are heterozygous at the *Rd* locus to greater sensitivity to environmental stresses causing more greening of the fruit. Interaction between virus infection and genes controlling fruit color may account for other cases of greening around the blossom end.

Clear differences between groups were observed for classic virus symptoms as shown by the AUDPC scores in table 3. GD (susceptible) had the highest level of infection (148.3), followed by the GD crosses (susceptible x resistant; mean = 71.2), and then by the resistant x resistant crosses (mean = 1.9), and these differences were statistically significant ($P < 0.0003$).

We conclude that green x orange fruit color crosses produce F1 hybrids that would not be acceptable to the processing industry because fruit, while mostly orange in color, have significantly more green around the blossom end. Unexpectedly some orange x orange crosses produced hybrid progeny with significant amounts of green at the blossom end, a result that suggests an interaction between color genes and ZYMV symptom expression. Resistance to ZYMV shows partial dominance, with resistant x susceptible crosses being intermediate to GD and resistant x resistant crosses. To achieve the highest levels of resistance with the desired skin color, it will be necessary for both inbreds to be orange-skinned and resistant. The current focus of our program is to backcross orange skin into the large green-fruited virus-resistant inbreds.

Literature cited

1. Brown, R.N. 2001. Traditional and molecular approaches to zucchini yellow mosaic virus resistance in *Cucurbita*. Ph.D. Thesis, Oregon State University, Corvallis, OR.

2. Lotsy, J.P. 1920. *Cucurbita* strijdvragen. II. Eigen onderzoekingen. Genetica 2:1-21.

3. McReynolds, R.B. 1996. Viruses affecting summer squash, cucumbers, and winter squash in the Willamette Valley. Annual Report. (unpublished).

4. Paris, H.S., and R.N. Brown. 2005. The genes of pumpkin and squash. Hortscience 40:1620-1630.

5. Shifriss, C. 1987. Studies on pigmentation of fruits in derivatives of crosses between *C. maxima* Duch. X *C. moschata* Poir. Plant Breeding 98:156-160

6. Shifriss, O. 1989. Relationship between the B genes of two *Cucurbita* species, II. Cucurbit Genet. Coop. Rpt. 12:75-78.

Table 1. Inbreds and OP cultivar used to produce F_1 hybrids evaluated in a trial planted at Corvallis, OR in 2008.

No.	Inbred	Skin color	No.	Inbred	Skin color
1	X1-1-2-1-1	Dark-green	12	X15-1-2-2-6	Red-orange
2	X1-1-2-1-2	Dark-green	13	X15-1-2-2-9	Red-orange
3	X1-1-2-1-3	Dark-green	14	X33-1-9-2-1	Red-orange
4	X7-2-1-3-1	Red-orange	15	X33-1-9-2-5	Red-orange
5	X7-2-1-3-4	Red-orange	16	X41-1-1-1-2	Red-orange
6	X7-2-1-4-2	Red-orange	17	X41-1-1-1-3	Red-orange
7	X7-2-3-1-2	Red-orange	18	X41-1-1-3-2	Red-orange
8	X7-2-3-1-4	Red-orange	19	X41-1-1-4-1	Red-orange
9	X7-2-3-3-5	Red-orange	20	X41-2-1-2-3	Red-orange
10	X15-1-2-2-3	Red-orange	21	X41-2-1-4-4	Red-orange
11	X15-1-2-2-5	Red-orange		Golden Delicious	Red-orange

Table 2. Yield of Golden Delicious derived squash hybrids grown under severe ZYMV infection at Corvallis, Oregon, 2008

	Marketable Fruit				Culls[z]	
Pedigree	No. ha^{-1}	MT ha^{-1}	Average Fruit Weight (kg)	Largest Fruit Weight (kg)	No. ha^{-1}	MT ha^{-1}
Golden Delicious	149	3.8	3.6	7.4	743	16.2
GD x X7-2-1-3-1	475	12.5	4.6	7.0	178	4.9
GD x X7-2-1-3-4	624	22.7	5.7	10.0	178	4.4
GD x X7-2-1-4-2	594	26.8	6.2	9.6	297	6.9
GD x X7-2-3-1-2	713	25.2	5.5	7.7	356	9.7
X15-1-2-2-3 x GD	921	21.6	3.5	7.4	297	6.0
GD x X15-1-2-2-6	743	22.3	4.6	7.5	238	5.3
X33-1-9-2-1 x GD	594	22.5	5.9	8.4	238	5.4
GD x X33-1-9-2-5	772	19.7	4.4	7.0	119	2.2
X41-1-1-1-2 x GD	1069	35.3	5.3	8.8	267	8.7
X41-1-1-3-2 x GD	683	20.7	3.9	6.8	535	7.9
GD x X41-1-1-4-1	832	24.9	4.5	8.6	267	5.7
X41-2-1-2-3 x GD	564	23.0	6.1	12.7	238	6.4
X1-1-2-1-2 x X7-2-1-3-1	416	25.7	9.7	17.3	446	24.4
X7-2-1-4-2 x X1-1-2-1-1	653	41.0	9.2	13.2	386	17.9
X1-1-2-1-2 x X7-2-3-1-4	564	37.8	9.4	13.2	446	20.2
X7-2-3-3-5 x X1-1-2-1-3	446	33.7	10.0	16.6	505	25.8
X1-1-2-1-2 x X15-1-2-2-5	446	24.7	7.9	11.9	416	16.8
X15-1-2-2-9 x X1-1-2-1-3	653	31.4	7.3	11.8	653	26.7
X33-1-9-2-5 x X1-1-2-1-3	1040	51.4	8.0	11.4	267	12.3
X41-1-1-1-3 x X1-1-2-1-2	1247	44.9	5.6	9.2	327	7.3
X41-1-1-3-2 x X1-1-2-1-1	594	27.3	7.1	12.2	356	13.7
X41-1-1-4-1 x X1-1-2-1-2	921	39.0	5.8	9.3	386	6.4
X41-2-1-2-3 x X1-1-2-1-1	653	35.3	7.7	11.8	802	33.2
X1-1-2-1-3 x X41-2-1-4-4	861	35.4	6.2	15.2	416	12.4
$LSD_{0.05}$	345	15.9	1.8		321	11.9

[z]Culls included immature and virus symptomatic fruit.

Table 3. Field Notes and Infection Scores for Golden Delicious Derived Winter Squash Lines, Corvallis, Oregon, 2008

Pedigree	Habit	Fruit Color	Green Blossom Ends	AUDPC Scores[z]	Powdery Mildew[y]
Golden Delicious	vine	red orange	none to slight	148.3	2.3
GD x X7-2-1-3-1	semi-bush	red orange	none to slight	112.3	5.3
GD x X7-2-1-3-4	semi-bush	pale red orange	none to slight	106.3	6.0
GD x X7-2-1-4-2	vine/semi-vine	red orange	none to slight	16.0	5.7
GD x X7-2-3-1-2	semi-bush	red orange	slight	54.7	5.7
X15-1-2-2-3 x GD	vine	red orange	slight	63.7	3.0
GD x X15-1-2-2-6	vine/semi-vine	red orange	large	82.7	3.0
X33-1-9-2-1 x GD	vine/semi-vine	red orange	none to slight	91.0	3.3
GD x X33-1-9-2-5	vine	red orange	none to slight	117.0	3.0
X41-1-1-1-2 x GD	vine/semi-vine	red orange	none to slight	32.0	4.3
X41-1-1-3-2 x GD	vine	red orange	none to slight	10.7	5.0
GD x X41-1-1-4-1	vine	red orange	none to slight	99.3	4.0
X41-2-1-2-3 x GD	vine	red orange	slight	68.3	5.0
X1-1-2-1-2 x X7-2-1-3-1	bush/semi-bush	pink orange	none to slight	8.3	5.3
X7-2-1-4-2 x X1-1-2-1-1	bush	pink orange	large	6.0	5.7
X1-1-2-1-2 x X7-2-3-1-4	bush/semi-bush	red orange	large	1.1	6.7
X7-2-3-3-5 x X1-1-2-1-3	bush/semi-bush	red orange	large	0.0	6.0
X1-1-2-1-2 x X15-1-2-2-5	semi-bush	red orange	large	0.0	3.7
X15-1-2-2-9 x X1-1-2-1-3	semi-bush	red orange	large	0.0	3.0
X33-1-9-2-5 x X1-1-2-1-3	vine	red orange	large	0.0	2.0
X41-1-1-1-3 x X1-1-2-1-2	vine	red orange	none to slight	2.3	5.0
X41-1-1-3-2 x X1-1-2-1-1	vine	red orange	none to slight	0.0	5.7
X41-1-1-4-1 x X1-1-2-1-2	vine	red orange	slight	0.0	5.7
X41-2-1-2-3 x X1-1-2-1-1	vine	red orange	large	5.3	5.3
X1-1-2-1-3 x X41-2-1-4-4	vine/semi-vine	red orange	none to slight	0.0	2.7
$LSD_{0.05}$				42.9	3.0

[z]Area Under the Disease Progression Curve score calculated by visually rating the plots three times with reading taken two weeks apart. Original data taken on a 1-5 scale where the number is number of plants in the plot that showed visual virus symptoms in either leaves or fruit. Maximum possible AUDPC score is 160. [y]Scale of 1-9; 9 = severe.

Figure 1. 'Golden Delicious' (*C. maxima*) fruit from a field trial conducted at the Lewis Brown Farm in Corvallis, OR in 2008 showing symptoms of zucchini yellow mosaic virus.

Figure 2. *C. maxima* hybrids X1-1-2-1-2 x X7-2-3-1-4 (green x orange, left) and GD x X7-2-1-2 (orange x orange, right) from a ZYMV infected field trial in Corvallis, OR in 2008 showing differences in the size of the green blossom end of the fruit. Golden Delicious (not shown) has a faint green ring around the stylar scar.

Confirmation of a Dominant Hard Rind (*Hr*) Locus in a *Cucurbita argyrosperma* ssp. *sororia* x *C. moschata* Cross

Linda Wessel-Beaver

Department of Crops & Agroenvironmental Sciences, Call Box 9000, University of Puerto Rico, Mayaguez, Puerto Rico 00681-9000 (lindawessel.beaver@upr.edu)

Introduction: One of the traits that often characterize fruits of non-domesticated *Cucurbita* is lignified (hard) rinds. Fruit of domesticated *Cucurbita* can have either lignified or nonlignified (soft) rinds. Schaffer et al. (6) noted that many summer squash cultivars of *C. pepo,* which are usually consumed within 10 days post-anthesis, have lignified rinds at maturity. On the other hand, winter squash cultivars, which need to be cut open at maturity for consumption, often are not lignified. Inheritance of lignification in *Cucurbita* was first studied by Mains (2) in both gourd and cultivated types of *C. pepo*. He concluded that this trait is controlled by a single dominant gene which was later given the symbol *Hr* by Robinson et al. (5). Schaffer et al. (6) confirmed that this is a single dominant trait in *C. pepo*.

In *C. maxima* control of rind lignification is more complex. Herrington and Brown (1988, as cited by Kock and Della Vecchia[1] and Paris and Brown [3]) studied rind lignification in crosses between soft rind *C. maxima* cv. Queensland Blue x hard rind *C. ecuadorensis* and noted that F1 progeny had soft rinds. They concluded that a dominant gene, *Hard rind inhibitor* (*Hi*), inhibits formation of a hard (lignified) rind in the presence of what is presumably the *Hr* gene in *C. maxima*. The hard rind of *C. ecuadorensis* presumably resulted from a *HrHr/hihi* genotype. Koch and Della Vecchia (1) carried out crosses between a hard rind *C. maxima* line and various soft rind *C. moschata* cultivars. All F1 progeny had hard rinds. In contrast, in crosses between the same hard rind *C. maxima* line and soft rind *C. maxima* cultivars, all F1 progeny had soft rinds. F2 progeny of these same crosses segregated 3:1 for soft:hard rind. Thus, Koch and Della Vecchia's study (1) suggests that many *C. maxima* cultivars are *HrHr* and are thus capable of producing lignified rinds, but instead have soft rinds because they are *HiHi* at the *Hard* rind *inhibitor* locus. The Japanese *C. moschata* cultivars used in the Koch and Della Vecchia (1) study were either *hrhr/hihi* or were *hrhr* without a corresponding *Hard* rind *inhibitor* locus (due to lack of complete homology between chromosomes of the two species). Whether the *Hi* locus is present in *Cucurbita* other than *C. maxima* needs further study.

In *C. moschata* and the closely related wild species *C. argyrosperma* ssp. *sororia,* previous work by this author and colleagues (Piperno et al. [4]) found that rind lignification is controlled by a single dominant gene *Hr* (*Hard rind*) in crosses between these species. However, this 2002 study (4) was primarily focused on the association between production of phytoliths and the hard rind trait (the association was confirmed). Phytoliths are solid particles of silicon dioxide found within the cells of *Cucurbita* species. The presence of phytoliths in soil samples can be used as a diagnostic tool in archeological studies of this genus. However, only small numbers of progeny were studied in the two F2 populations included in the study: 20 individuals in one population and 30 individuals in the other population. No reciprocal crosses were made. Therefore, the purpose of this study was to confirm the monogenic dominant inheritance of rind lignification in *C. sororia* x *C. moschata* with additional reciprocal backcross data.

Materials and Methods: *C. argyrosperma* ssp. *sororia* (Sor) accession Sor177 (of Mexican origin) was crossed with pollen parent *C. moschata* ´PRShortvine´ (Mos) (an experimental line from the University of Puerto Rico, Mayaguez breeding program) to produce F1 seed. A single F1 plant was backcrossed to *C. moschata* ´PRShortvine´, using the latter as both the pollen and seed parent (reciprocal backcrosses). Backcross seed was planted at Isabela, Puerto Rico. Several plants of each parent were also grown at the same time. A single fruit was harvested from each of 88 F1 x Mos progeny and 65 Mos x F1 progeny. A small rind sample (about 3 cm x 8 cm, with flesh removed) was taken from each fruit and oven dried.

Results and Discussion: When progenies were classified on the basis of rind lignification, both backcross populations segregated 1:1 (lignified:non-lignified) (Table 1). Samples of non-lignified rinds remained soft and somewhat leathery, and curled as they dried. Lignified rinds retained most of their shape (minimum curling) and were very hard once dried. Thickness of lignified rinds in the *C. sororia* parent averaged about 1.5 mm. In backcross progeny thickness of lignified rinds varied from 1 mm to as much as 3 mm. For unknown reasons progenies from the Mos x F1 backcross had a larger number of very thick rinds compared to the F1 x Mos backcross (data not shown). The range in rind thickness in the backcross populations suggests that there may be other genes modifying deposition of lignin in the rind. The author has tested a large number of genotypes of both *C. sororia* and *C. moschata* in crosses and has always observed hard (lignified) rind to be dominant in the F1 in contrast to what Herrington and Brown (1988, as cited by Kock and Della Vecchia[1] and Paris and Brown [3]) observed in *C. maxima* x *C. ecuadorensis.* The author has also tested F1 progeny of lignified x non-lignified *C. moschata* and observed these fruit to always be lignified. These observations along with the data presented here confirmed rind lignfication (*Hard rind, Hr*) to be a single dominant trait in *C. moschata* and *C. argyrosperma* ssp. *sororia.* These observations also suggest that the *Hi* (*Hard* rind *inhibitor*) locus does not occur in *C. moschata* and *C. argyrosperma* ssp. *sororia* or that most or all genotypes of these species carry both recessive alleles (*hihi*).

Acknowledgement: The author wishes to thank Thomas C. Andres for providing the original seed stock of Sor177.

Literature Cited

1.Koch, P.S. and P.T. Della Vecchia. 1994. Inheritance of hard rind in *C. maxima* x *C. moschata* Crosses. Cucurbit Genetics Cooperative Report 17:122-124.

2.Mains, E. B. 1950. Inheritance in *Cucurbita pepo*. Papers of the Michigan Academy of Science, Arts and Letters 36: 27-31.

3.Paris, H. S. and R. N. Brown. 2005. The genes of pumpkin and squash. HortScience 40(6):1620-1630.

4.Piperno, D. R., I. Holst, L. Wessel-Beaver and T.C. Andres. 2002. Evidence for the control of phytolith formation in *Cucurbita* fruits by the hard rind (*Hr*) genetic locus: archaeological and ecological implications. PNAS 99(16):10923-10928.

5.Robinson, R.W., H.M. Munger, T.W. Whitaker and G.W. Bohn. 1976. Genes of the Cucurbitaceae. HortScience 11:554-568.

6.Schaffer, A.A., Ch. D. Boyer and H.S. Paris. 1986. Inheritance of rind lignification and warts in *Cucurbita pepo* L. and a role for phenylalanine ammonia lyase in their control. Z. Pfanzenzüchtg 96:147-153.

Table 1. Observed and expected numbers of lignified and non-lignified fruits in wild (Sor 177 = *Cucurbita argyrosperma* ssp. *sororia*) and domesticated (Mos = *C. moschata* ´PRShortvine´) *Cucurbita* and their reciprocal backcross progeny.

Population	Lignified rind	Non-lignified rind	Expected ratio	Chi-square	Prob
Sor 177 (Sor)	5	0	1:0		
Mos	0	6	0:1		
(Sor x Mos) x Mos	39	49	1:1	1.14	0.286
Mos x (Sor x Mos)	26	39	1:1	2.60	0.107

An Austrian Cucumber Mosaic Virus Isolate is Causing Severe Symptoms on Resistant *Cucurbita pepo* Cultigens

Martin Pachner[1] and Tamas Lelley[2]

[1] University of Natural Resources and Applied Life Sciences, Vienna

[2] Department for Agrobiotechnology, Institute of Biotechnology in Plant Production, IFA-Tulln, Konrad Lorenz Str. 20, A-3034 Tulln, Austria <tamas.lelley@boku.ac.at>

This article is being reprinted in its completion, from 2007.

Introduction: In 2005, we discovered some plants of a zucchini yellow mosaic virus (ZYMV)-tolerant oil-pumpkin breeding line (*Cucurbita pepo*), severely affected by a virus. Our first assumption was that ZYMV might have overcome the resistance, but ELISA tests revealed that most likely cucumber mosaic virus (CMV) caused the symptoms, although ZYMV was detected in very low concentrations as well. Using fruit flesh of such infected plants for inoculation of pumpkin seedlings lead to immediate death of all plants, independent of whether the plants were ZYMV-tolerant or susceptible. Therefore, we decided to isolate the CMV for further investigation.

Materials and methods: *Artificial inoculation.* An Austrian isolate of CMV was established as follows. Fruit flesh from oil-pumpkin with multiple virus infestation was collected in fall 2005 and used to inoculate tobacco plants. Tobacco, *Nicotiana tabacum,* is not susceptible to zucchini yellow fleck potyvirus (ZYFV) and squash mosaic comovirus (SqMV) (Plant Viruses Online: http://image.fs.uidaho.edu/vide/famly124.htm#Nicotiana%20tabacum). Tobacco is also not known to be susceptible to ZYMV, the most common virus in oil-pumpkin. Leaves of tobacco plants showing severe mosaic were tested with ELISA for CMV, ZYMV, WMV2 and SqMV. The presence of any virus other than CMV was excluded. Then CMV was increased on plants of the susceptible *C. pepo* variety Gleisdorfer Ölkürbis, because inoculations on *Cucurbita moschata*, using infected leaves of tobacco, failed. The Hungarian isolate (HI), provided to us by István Tóbiás (Plant Protection Institute, Hung. Acad. Sci., Budapest, Hungary), was purified and tested in the same way as the Austrian isolate (AUTI). The French isolate (FI), received from Muriel Archipiano (Clause Tézier, Domaine de Maninet, Route de Beaumont, Valence, France) was treated in the same way. The inoculum for the experiments was prepared from 1.0 g of infected leaves, ground in a mortar on ice, in 10ml inoculation buffer containing 1% K_2HPO_4. Finally, 1.0 g Celite® 545 was added. Seedlings were inoculated twice: first the two cotyledons, when the first true-leaf just appeared, and three days later the first true-leaf itself. Inoculation was carried out by gently rubbing the leaf surface with a finger in rubber gloves. The leaves were rinsed with water immediately after rubbing. Simultaneously the Hungarian and French isolates of CMV were tested.

Plant material: Nine *C. pepo* and nine *C. moschata* cultigens (Table 1) were grown in pots in the greenhouse at 23°-25°C day and 20°-22°C night temperatures at 50-70% RH. Natural illumination was supplemented with a combination of mercury and sodium vapor lamps (ca. 10,000 lux), maintaining a day-length of 14 hours during the whole experiment.

Evaluation: Plants were observed 14 and 24 days after the first inoculation. Leaf symptoms (LS) were rated from 0 (no symptoms) to 9 (severe mosaic). A rating of 10 was introduced for dead plants. Additionally, the approximate growth reduction (GR), in relation to normal growth, was scored in percent. For further evaluation a total rating (TR), using the formula TR=LS+GR*0.05, was calculated. Plants with a TR=0 to 5 were classified as tolerant, such with TR>5 as susceptible. TR-values greater than 10 were limited to 10. To verify the TR-value, we applied the 0 to 5 rating system described by Walkey and Pink (4), who combined leaf symptoms and stunting in one score. After the last evaluation the experiment was terminated and the plant material was autoclaved.

Results and Discussion: A comparison of results obtained by the infection experiment (Table 1), shows that AUTI is the most aggressive isolate. The symptoms caused by HI were half as severe as those caused by AUTI, those caused by FI were still somewhat milder. Comparisons of results obtained with *C. pepo* and *C. moschata*, revealed that, except against AUTI, most of the *C. moschata* cultigens showed a high level of CMV resistance. 'Nigerian Local', however, developed severe symptoms when inoculated with AUTI, although we had hoped that it could be the source of a high level of resistance, as was reported by Brown et al. (1). Nigerian local was found to be resistant against a number of viruses and was therefore used in many breeding programs (1). We obtained a similar result with 'Menina 15' (received from Michael Pitrat, INRA, Montfavet, France), which is, analog to Nigerian Local, highly resistant against ZYMV (2). Only 'Zhou', a Chinese, hull-less *C. moschata* cultivar named by us according to its discoverer Zhou Xianglin (5) and Soler, (kindly provided by L. Wessel-Beaver, USDA-ARS, Puerto Rico), seemed to have resistance against AUTI (Fig. 1). All *C. pepo* cultigens, including 'Linda', an American zucchini F1 variety from Harris Moran Seed Company (Modesto, California) described as CMV-resistant, showed high susceptibility to AUTI. The zucchini variety True French (kindly provided by Harry Paris, Newe Ya`ar Res. Center, Ramat Yishay, Israel), developed clearly less leaf symptoms than most of the other *C. pepo* cultigens. 1997, for the first time, a ZYMV-epidemic destroyed half of the oil-pumpkin harvest in Austria (3). We are alarmed by the fact that, in our first experiment, CMV in combination with ZYMV killed all our test plants. We are wondering, why CMV in the field so far occurs only on single plants. One possibility could be that AUTI lost its aphid transmissibility. A sequencing of the virus genome is in progress. Further investigations will have to be carried out to determine the potential danger posed by this isolate.

Literature Cited

1.Brown, R. N., A. Bolanos-Herrera, J. R. Myers, and M. M. Jahn. 2003. Inheritance of resistance to four cucurbit viruses in *Cucurbita moschata*, Euphytica 129:253–258.

2.Lecoq, H., M. Pitrat, and M. Clement. 1981. Identification et caracterisation dún potyvirus provoquant la maladie du rabougrissement jaune du melon. Agronomie 1:827-834.

3.Riedle-Bauer, M. 1998. Ölkürbis & Co.: Was tun gegen das Zucchinigelbmosaikvirus? Der Pflanzenarzt 51:27-30.

4.Walkey, D. G. A., and D. A. C. Pink. 1984. Resistance in vegetable marrow and other *Cucurbita* spp. to two British strains of cucumber mosaic virus, J. agric. Sci. Camb. 102:197-205.

5.Xianglin, Z. 1987. A study on the breeding of naked kernel pumpkin and its genetic behaviour. Acta Hort. Sinica 14:114-118. (in Chinese with English summary)

Table 1: Tested cultigens, their geographic origin, resistance behaviour, average rating for leaf symptoms (LS), growth reduction in % to control (GR), calculated total rating (TR) and Walkey and Pink (4) rating (WP) for comparison.

Cultigens/Origin	Resistance	CMV-AUTI[y]				CMV-HI[x]				CMV-FI[w]			
		LS	GR	TR	WP	LS	GR	TR	WP	LS	GR	TR	WP
		C. pepo											
True French Zucchini/GB	no	3.0	50	5.5	4.0	2.0	25	3.3	3.0	2.8	40	4.8	3.6
Gleisdorfer Ölkürbis/AUT	no	8.7	96	9.3	4.8	4.3	33	6.0	3.3	3.3	29	4.8	3.2
Linda Zucchini F1/USA	CMV	5.0	53	7.6	4.0	3.5	21	4.5	2.7	2.3	17	3.2	2.5
Linda selfing	CMV[z]	5.3	54	7.9	4.0	3.7	58	6.6	3.8	2.8	50	5.3	3.7
Linda x Gleisdorfer Ölkürbis		5.5	63	7.8	4.0	4.5	42	6.6	3.7	3.0	25	4.3	3.0
Tigress Zucchini F1/USA	ZYMV	5.2	50	7.7	4.0	4.5	25	5.8	3.0	3.8	40	5.8	3.6
Tigress selfing	ZYMV[z]	4.7	58	7.6	3.8	4.8	33	6.5	3.3	4.0	63	7.1	3.8
44/15 oil-pumpkin breeding line/AUT	ZYMV	5.7	67	7.3	4.2	4.3	42	6.4	3.7	3.0	29	4.5	3.2
True French resistant against ZYMV/Israel	ZYMV	9.2	79	10.0	5.0	2.0	75	5.8	4.0	3.2	50	5.7	4.0
Mean		6.0	65	8.0	4.2	3.7	39	5.7	3.4	3.1	38	5.0	3.4
		C. moschata											
Waltham Butternut/USA	no	4.2	75	7.3	4.2	4.0	45	6.3	3.8	0.5	0	0.5	0.5
Zhou x Watlham Butternut.		3.0	25	4.3	3.0	0.0	8	0.4	0.7	0.2	4	0.4	0.5
Zhou hull-less/China		1.4	25	2.7	3.0	0.8	0	0.8	0.5	0.0	0	0.0	0.0
Soler/Puerto Rico	ZYMV	2.2	25	3.4	3.0	0.2	4	0.4	0.5	0.0	0	0.0	0.0
Waltham Butternut. x Soler		3.0	50	5.5	4.0	1.3	0	1.3	1.3	0.3	0	0.3	0.3
Menina 15/France	ZYMV	5.0	50	7.5	4.0	0.0	0	0.0	0.0	0.0	0	0.0	0.0
Nicklow´s Delight selfing/USA		4.2	79	7.3	4.2	3.0	25	4.3	3.0	2.0	4	2.2	2.2
Nigerian Local/Nigeria	ZYMV, CMV, WMV2, PRSV-W	4.0	75	7.8	4.0	0.0	0	0.0	0.0	0.0	0	0.0	0.0
Waltham Butternut x NigerianLocal		5.0	50	7.5	4.0	0.5	4	0.7	0.5	0.0	0	0.0	0.0
Mean		3.7	52	6.0	3.7	1.1	10	1.6	1.1	0.3	1	0.4	0.4

[z] Segregating; [y] Austrian isolate; [x] Hungarian isolate; [w] French isolate

Fig. 1 Symptom development 24 days after inoculation with three isolates of CMV on the C. *moschata* cultigens 'Waltham Butternut', 'Zhou', and 'Nigerian Local'

The Quality of Seed of *Cucurbita* sp. is Determined by the Development of its Embryo

J.R. Gardingo, T. Santos, A. Barabach and R.R. Matiello
Departamento de Fitotecnia e Fitossanidade, Universidade Estadual de Ponta Grossa, Av Carlos Cavalcanti, 4748, Uvaranas, CEP 84030-900, Ponta Grossa, Paraná, Brasil. e-mail: jrgardin@uepg.br

Introduction: The seeds of *Cucurbita maxima*, *C. moschata* and *C. pepo* present an excellent potential for human nourishment since they are a source of proteins (1) and minerals. Their lack of use is due to the hardness of the hull and lack of knowledge regarding the existence of hull-less seeds of *C. pepo* and *C. maxima*. In Brazil, the seeds are used mainly as the component of the flour of a mix of grains to avoid infantile malnutrition.

The use of the seed for the multiplication of the species requires that the seed be of high quality, that is, presents a well-developed embryo. Such seeds, when germinated, result in vigorous and healthy seedlings, essential for long-term germplasm storage. Most publications relate seed germination of *Cucurbita* sp. with its biomass, age and/or fruit storage period, with few reports assessing the specific development of the embryo (2,4,5). The development of the embryo depends on several factors such as the sowing location and the genotype. The embryos of *C. pepo*, *C. maxima* and *C. moschata* become visible to the naked eye between 20 and 25 days post anthesis (PA), and their biomass reaches the maximum level between 50 and 60 days PA (2,3,4). The development of the embryo can be changed by fruit age and by storage time after harvest, where this last factor results, generally, in an increase of the biomass of the embryo (2,3,4).

Materials and methods: The cultivars ´Menina Brasileira Precoce´ (*C. moschata*) and ´Exposição´ (*C. maxima*) were planted on 17 October 2007 and 18 January 2008. The plants were cultivated at *Fazenda Escola Capão da Onça* (Capão da Onça Farm School), *Universidade Estadual de Ponta Grossa* (State Univeristy of Ponta Grossa), Ponta Grossa (25° S; 50° 10′ W), Paraná State, Brazil. The spacing between the rows was 4 m and 3 m between the plants in the row. Chemical control was employed against insects and diseases. The flowers were protected by paper bags after being manually pollinated. During the month of January, 8 fruits were harvested with the ages of 5, 10, 15, 20, 25, 30, 35, 40, 45, 50, 55, 60, and 65 days PA. The length through the axis and the maximal width of seeds were measured. Samples of the fruits harvested at the age of 15, 30, 45, and 60 days PA were stored up to 60 days. The seeds were removed from the fruits with or without storage. During the month of May, fruits of only some ages were harvested. Ten seeds and respective embryos were used to obtain the largest lengths and widths. The fresh biomass was measured from 50 seeds and 10 embryos.

Results and Discussion

The seeds of *C. moschata* (Table 1) and *C. maxima* (Table 2) expanded rapidly during the fruit expansion. The hull of the seed became rigid while the fruit remained connected to the plant. The hull was the main component of biomass of the seed until 35 PA (4). Since the length and the maximum width of the seed are achieved at 15 days PA, such characteristics are not useful for the determination of quality of the embryo.

Although the seed appears developed in fruits up to 30 days PA, corroborating the results of Loy (2), the embryo of *C. moschata* or *C. maxima* becomes visible to the naked eye only between 20 and 25 days PA. After the embryo achieved 2 mm of length, rapid increases of length and width were noticed until 40 days PA for *C moschata* (Table 1) and 35 days PA for *C. maxima* (Table 2). After this period, the increase of the width of the embryos resulted in wider seeds when dehydrated. The presence of gelatinous endosperm around the embryo is a characteristic of immaturity of the embryo. The presence of gelatinous endosperm occurs in seeds of *C moschata* between 20 and 45 days PA and seeds of *C. maxima* between 20 and 35 days PA. Thus, it is possible to affirm that ´Exposição´ is more precocious compared to the ´Menina Brasileira Precoce´.

The climatic conditions during plant growth can affect the development of the embryo (2). The fruits of *C. moschata* (Table 1) and *C. maxima* (Table 2) harvested during the month of May (autumn) presented a development inferior to those harvested during the month of January (summer). Besides the embryo, the formation of the gelatinous endosperm was delayed but had its amount increased for *C moschata* (30 and 65 days PA) and for *C. maxima* (25 and 45 days PA). Such delay in the development of the embryo occurred due to the reduction of the temperature and luminosity.

Regardless of the amount of time the fruit is attached to the plant, seed quality can be improved by the storage of the fruits, resulting in the increase of biomass of the embryo (2,3,4). The fruits of *C. moschata* and *C. maxima* harvested 15 days PA presented an increase in the length and width of the embryo during storage time (up to 45 days). Regardless of the storage period used for these fruits, the amount of the seeds and size of the embryos were inferior to those obtained from fruits harvested 30, 45, and 60 days PA under the same conditions. Therefore, the harvest of the fruits 15 days PA for the obtention of seeds is not recommended. The storage of the fruits harvested 15, 30, 45, and 60 days PA resulted in the increase of biomass, benefiting the embryo. The measurement of the length and width of some embryos per fruit is a rapid method for assessing the stage of seed development. The determination of the quality of seed can become more accurate using as auxiliary parameters the determination of the width and dry biomass of the embryo.

Literature Cited

1. Jacks, T.J., T.P. Hensarling and L.Y. Yatsu. 1972. Cucurbit seeds: I. Characterizations and uses of oils and proteins. A Review. Econ. Bot. 26: 135-141.
2. Loy, J.B. 2000. Seed development in *Cucurbita pepo*: An overview with emphasis on hull-less seeded genotypes of pumpkin. Cucurbit Genetics Cooperative Report, 23: 89-95.
3. Santos, T. 2008. Desempenho da qualidade de sementes de abóbora cv Menina Brasileira colhidas em quatro épocas e submetidas a diferentes períodos de armazenamento. 34p. Trabalho de Conclusão de Curso Agronomia. UEPG www.uepg.br/colegiados/colagro/monografias/TatianeDosSantos.PDF.
4. Barabach, A. 2009. Qualidade de sementes de moranga 'Exposição' em função da idade e do armazenamento de frutos. 2009. Trabalho de Conclusão de Curso Agronomia. UEPG. 42p.
5. Nakajima, T. 1962. Physiological studies of seed development, especially embryonic growth and endosperm development. Bull. Univ. Osaka Pref., Ser. B, 13: 13-48.

Table 1. Length and width of seeds of embryos of *Cucurbita moschata* 'Menina Brasileira Precoce' obtained from fruits harvested in January (Summer) and May (Autumn) - 2008. Ponta Grossa, Paraná State, Brazil.

	Summer				Autumn			
	Seeds		Embryos		Seeds		Embryos	
Days post-anthesis	Length[1] mm	Width[2] mm	Length mm	Width mm	Length mm	Width mm	Length mm	Width mm
15	17.3	11.6	*	*	16.7	8.5	*	*
20	14.7	10.0	4.5	3.0	14.5	9.3	*	*
25	-	-	-	-	-	-	-	-
30	14.6	9.9	6.1	4.5	15.6	10.0	3.9	2.7
35	14.6	10.0	9.2	4.9	-	-	-	-
40	14.4	10.2	10.2	7.5	-	-	-	-
45	14.3	10.3	10.5	7.8	-	-	-	-
50	18.6	11.9	11.6	8.1	-	-	-	-
55	16.0	10.3	13.0	8.6	15.4	10.1	10.6	7.6
60	16.6	10.7	13.0	8.9	-	-	-	-
65	15.9	10.9	12.1	8.8	22.3	13.9	10.9	7.7

[1] Length of recently harvested seeds (mm),
[2] Width of recently harvested seeds (mm)
* not visualized by the naked eye
\- not determined

Table 2. Length and width of seeds of embryos of *Cucurbita maxima* 'Exposição' obtained from fruits harvested in January (Summer) and May (Autumn) - 2008. Ponta Grossa, Paraná State, Brazil.

	Summer				Autumn			
	Seeds		Embryos		Seeds		Embryos	
Days post-anthesis	Length[1] mm	Width[2] mm	Length mm	Width mm	Length mm	Width mm	length mm	Width mm
5	6.3	3.5	*	*	-	-	-	-
10	14.1	9.9	*	*	-	-	-	-
15	16.6	11.4	*	*	-	-	-	-
20	15.7	11.1	1.8	1.3	16.1	10.8	-	-
25	16.5	11.3	8.9	5.8	16.7	10.6	4.1	2.2
30	16.2	11.5	13.1	8.9	15.7	9.8	4.8	2.8
35	16.8	11.5	13.9	9.6	-	-	-	-
40	16.6	11.2	14.2	9.6	-	-	-	-
45	15.7	11.0	14.2	9.8	-	-	-	-
50	16.9	11.3	14.4	9.7	17.2	9.7	14.2	7.3
55	16.0	10.3	14.4	9.3	-	-	-	-
60	16.0	10.8	14.0	10.0	15.2	10.5	12.8	8.9

[1] Length of recently harvested seeds (mm),
[2] Width of recently harvested seeds (mm)
* not visualized in the naked eye
- not determined

Evaluation of Sponge Gourd Hybrids for Yield and Related Traits

Sabina Islam*[1], A. D. Munshi[1], Ravinder Kumar[1], T. K. Behera[1] and S.K. Lal[2]
[1]Division of Vegetable Science
[2] Division of Genetics
Indian Agricultural Research Institute, New Delhi, 1100012
** Email address of corresponding author: <tokumkum@gmail.com>*

Abstract: Ten F_1 hybrids of sponge gourd developed through Line × Tester method were evaluated for yield and related traits. Significant variation in mean performance was noticed for all the characters studied. Different hybrids were found best for different traits. The best performing hybrids for different characters include DSG-6 × 'Pusa Sneha' for earliness; DSG-7 × PSG-9 for fruit length and average fruit weight; DSG-6 × CHSG-2 for fruit diameter and vine length; DSG-7 × NSG-1-11 for number of fruits per plant, average fruit weight and total yield per plant.

Spongegourd (*Luffa cylindrica* Roem.), is a very popular vegetable in the tropical and subtropical regions. It is an important component of crop rotation during spring-summer and rainy season in North Indian condition and is cultivated both on commercial scale and in kitchen gardens (Choudhury, 1996). The young tender fruits of the non-bitter types are eaten as cooked vegetable, or used in soups. The seed oil is colourless, odourless and tasteless which is used in cooking. The plants have medicinal properties too. Fiber is obtained from fully ripen and dried fruits which is useful in cleaning the motor car, glassware, kitchen utensils, commercial filters, for insulation in pot-holders, bathmats, and related uses (Porterfield, 1955). Despite its importance and diversified use, very little attention has been made for improvement in horticultural traits. Very few varieties and hybrids are grown for commercial cultivation in India. Being monoecious and essentially cross pollinated, it provides ample scope for successful exploitation of hybrid vigour. Hence the present study was undertaken to evaluate the performance of hybrids of sponge gourd along with their parents for yield and related traits.

Material and methods

Seven diverse lines obtained through National Bureau of Plant Genetic Resources, New Delhi were maintained as inbreds at Division of Vegetable Science, Indian Agricultural Research Institute, New Delhi. These inbreds were crossed in Line × Tester mating design to obtain ten F_1 hybrids and these hybrids along with seven parents were evaluated for yield and yield related traits. The experiment was laid out in randomized block design with three replications. Plant to plant distance was maintained 80 cm and rows were made at 120 cm apart. A single non-experimental row was planted on either sides of each block so as to minimize environmental error due to border effect. All the recommended cultural practices were followed to raise a healthy crop. Data were recorded on ten randomly selected plants in each treatment (hybrids and parents) for seven characters *viz.* vine length, days to first fruit picking, number of fruits per plant, fruit length, fruit diameter, average fruit weight and total yield per plant.

Results and Discussion

The mean performances of parents (inbreds) and hybrids for various traits have been presented in Table 1. The analysis of variance (data not presented) was carried out to test the significance of differences among parents and their hybrids. This result clearly indicated that there were significant variations in mean performance among parents and their hybrids for all the characters studied. A perusal on average performance of parents and hybrids (Table 1) revealed that the mean values of parents for vine length ranged from 4.0 m ('Pusa Sneha') to 6.4 m (DSG-7) whereas for hybrids it ranged from 4.5 m to 6.8 m and the hybrid, DSG-6 × CHSG-2 was found to be the best. The mean for days to first female flowering ranged from 53.67 (DSG-6) to 65.33 days (CHSG-2) among the parents, whereas the hybrid, DSG-6 × 'Pusa Sneha' took least time (50.0 days) to produce marketable fruits, and DSG-6 × PSG-9 took maximum time (64.0 days; Table 1). The number of fruits per plant varied from 5.1 (PSG-9) to 22.6 (DSG-6) in parents and among the F_1 hybrids, DSG-7 × NSG-1-11 produced the highest average number of fruits per plant (29.4) and DSG-6 × CHSG-2 produced the least (9.9). The mean value of fruit length varied from 13.4 (DSG-6) to 14.2 cm

('Pusa Sneha') for parental lines. The F_1 hybrid, DSG-7 × PSG-9 showed the highest fruit length (20.3 cm) and DSG-7 × CHSG-1 produced the shortest fruit (12.0 cm). The fruit diameter varied from 3.5 cm (DSG-7 × CHSG-2) to 4.9 cm (DSG-6 × CHSG-2) while comparing both parental lines and F_1 hybrids. The average fruit weight ranged from 96.3 g (CHSG-1 and CHSG-2) to 126.0 g (DSG-6), while in F_1 hybrids it varied from 118.0 g (DSG-7 × CHSG-1) to 154.0 g (DSG-6 × PSG-9 and DSG-7 × NSG-1-11). Yield/plant is considered the most important character for any crop improvement. In the present study among the parents, DSG-6 produced highest fruit yield per plant (2.85 kg) and PSG-9 produced the least fruit yield per plant (0.50 kg) while among crosses, the total yield per plant was maximum in DSG-7 × NSG-1-11and least in DSG-7 × CHSG-2 (1.28 kg). Based on *per se* performance of the F_1 hybrids, DSG-6 × 'Pusa Sneha' was found to produce early yield and DSG-7 × NSG-1-11 was the most high yielding hybrids. The earliness may be attributed to minimum days to male and female flowering, lower node of flowering, while increased yield was due to increase in fruit size, weight and number of fruits per plant. These two most promising hybrids can be recommended for commercial cultivation in North Indian plains to benefit the growers to catch the early market as well as increased yield. Similar reports of superior performance of hybrids were reported by Shaha and Kale (2003) and Hedau and Sirohi (2004) in ridge gourd and Lou *et al.* (2005) in angular sponge gourd.

Literature Cited

Choudhury, B. 1996. Vegetables. National Book Trust, New Delhi. p.168.

Hedau, N.K. and Sirohi, P.S. (2004). A diallel studies in ridge gourd [*Luffa acutangula* (Roxb) L.]. *Orissa Journal of Horticulture*. **32**(1): 13-14.

Lou, J.N., He, X.L. and Lou, S.B. (2005). A new sponge gourd F1 hybrid - 'Yueyousigua'. *China Vegetables*. **5**: 26-27.

Porterfield, W. M., (Jr.) (1955). Loofah - the sponge gourd. *Economic Botany*. **9**: 211-223.

Shaha, S.R. and Kale, P.N. (2003). Heterosis studies in ridge gourd. *Journal of Maharashtra Agricultural Universities*. **28**(3): 334-336.

Table 1. *Per se* performance of parents, their hybrids, and heterosis over better parent for yield and yield related traits

Parents / F_1	Vine length (m)	Days to first fruit picking	Number of fruits/plant	Fruit length (cm)	Fruit diameter(cm)	Fruit weight(g)	Total yield/plant (kg)
DSG-6	5.49	59	22.63	13.37	3.93	126	2.85
DSG-7	6.38	61.33	20.90	13.47	3.8	123.67	2.59
'Pusa Sneha'	4.00	60.67	18.67	14.17	3.67	117.33	2.19
PSG-9	5.74	62	5.13	13.67	3.67	96.33	0.50
NSG-1-11	5.79	62.67	6.5	14.27	3.6	96.33	0.63
CHSG-1	6.06	65	7.37	13.47	3.73	111	0.81
CHSG-2	5.89	65.33	8.23	13.87	3.67	108.67	0.89
DSG-6 × 'Pusa Sneha'	5.72 (4.19)	50.0 (-15.25**)	25.27 (11.67*)	18.23 (28.65**)	4.77 (21.37*)	148.0 (17.46**)	3.86 (31.93**)
DSG-6 × PSG-9	6.32 (10.10)	63.33 (7.34)	28.40 (25.50**)	18.0 (31.67**)	4.47 (13.74)	154.0 (22.22**)	4.38 (53.68**)
DSG-6 × NSG-1-11	5.45 (-5.87)	64.33 (9.03)	25.80 (14.01**)	16.47 (15.42)	4.37 (11.20)	152.67 (21.17**)	3.94 (38.25**)
DSG-6 × CHSG-1	4.71(-22.28**)	63.33 (7.34)	24.03 (6.19)	14.37 (6.68)	3.73 (-5.09)	144.33 (14.55**)	3.48 (22.11**)
DSG-6 × CHSG-2	6.81(15.62)	54.33 (-7.92)	27.63 (22.09**)	19.27 (38.93**)	4.87 (23.92*)	150.67 (19.58**)	4.17 (46.32**)
DSG-7 × 'Pusa Sneha'	5.42 (-15.05)	51.67 (-14.83**)	22.57 (7.99)	15.43 (8.89)	4.53 (19.21*)	143.67 (16.17**)	3.24 (25.10**)
DSG-7 × PSG-9	4.55 (-28.68**)	59.33 (-3.26)	18.83 (-9.90)	20.33 (48.72**)	3.70 (-2.63)	135.33 (9.43)	2.56 (-1.16)
DSG-7 × NSG-1-11	5.32 (-16.61*)	56.67 (-7.60)	29.43 (40.81**)	20.0 (40.15**)	4.60 (21.05*)	154.0 (24.52**)	4.53 (74.90**)
DSG-7 × CHSG-1	4.62 (-27.59*)	55.33 (-9.78*)	11.40 (-45.45**)	12.07 (-10.39)	3.57 (6.05)	118.0 (-4.58)	1.34 (-48.26**)
DSG-7 × CHSG-2	5.91 (-7.37)	51.0 (16.84**)	9.9 (-52.63**)	14.27 (2.88)	3.5 (-7.89)	128.33 (3.77)	1.28 (-50.58**)
Mean value of Parents	5.62	62.29	12.78	13.76	3.72	111.33	1.49
Mean value of hybrids	5.48	56.93	22.33	16.84	3.86	142.9	3.27
$SE_{(d)}$	0.52	2.71	1.02	1.23	0.35	7.01	0.23
$CD_{(0.05)}$	1.05	5.53	2.08	2.50	0.72	14.29	0.46
$CD_{(0.01)}$	1.41	7.43	2.80	3.36	0.96	19.21	0.62

( ): Percent heterosis over better parent is given; * Significant at 5% level; ** Significant at 1% level.

Length and Rapid Elongation of Pedicels of the Female Flowers of *Cucumis anguria* L.

Mark P. Widrlechner, Kathleen R. Reitsma, and Lucinda D. Clark
USDA-ARS North Central Regional Plant Introduction Station, Iowa State University, Ames, IA 50011-1170

Joseph H. Kirkbride, Jr.
USDA-ARS, U.S. National Arboretum, 3501 New York Avenue, NE, Washington, DC 20002-1958

Introduction: Our recent work describing *Cucumis zambianus* Widrlechner, J.H. Kirkbr., Ghebretinsae & K.R. Reitsma, a new species from Zambia, led us to spend considerable time documenting inflorescence characteristics in this new species and comparing them to other, similar *Cucumis* taxa (10). *Cucumis zambianus* and *C. anguria* share a trait that is rather unusual, pedicels that are often considerably longer than the fruits they subtend. However, in *C. zambianus*, the pedicels are of considerable length (65-120 mm) at the time that female flowers open (10), while *C. anguria* is reported by some authors (but not others) to have much shorter pedicels at that developmental stage, with elongation evidently occurring rapidly (8) during the course of fruit maturity. Kirkbride (7) indicated that the pedicels of female flowers of *C. anguria* are initially quite short, ranging from 1.5 to 7 mm, but other authors, including Howard (3) and Jeffrey (4-6) reported much longer pedicels, from 13 to 105 mm.

Reported values are much longer for the fruit pedicels and also more consistent among authors. The widest range in fruit-pedicel length was reported by Jeffrey (5) at 25 to 210 mm, with five other reports (2, 4, 6-8) all in agreement on lengths between 60 and 135 mm. This suggests that the pedicels of female flowers at least double in length after fertilization, and perhaps could lengthen by as much as 30×.

The magnitude of this growth seemed remarkable, leading us to design an experiment to measure the pedicels of female flowers from the time that the flowers open through fruit development for six accessions of *C. anguria* representing both botanical varieties, the typical variety and var. *longaculeatus* J. H. Kirkbr. This allows us to evaluate the discrepancy between the descriptions of Kirkbride (7), which were also used by Schaefer (9) in a recent re-examination of the genus *Cucumis*, and those of other authors (3-6) and to document the magnitude and rate of pedicel elongation.

Materials and Methods: Six accessions of *Cucumis anguria* (Table 1) were selected from the *Cucumis* collection maintained at the USDA-ARS North Central Regional Plant Introduction Station in Ames, Iowa, USA. The accessions included two *C. anguria* var. *anguria* and four *C. anguria* var. *longaculeatus*, chosen for broad geographic representation.

To overcome dormancy, seed coats were chipped prior to planting. Eight seeds were planted on 10 February 2009 in each of three, 30 cm, round plastic pots per accession. Seedlings were thinned at the second true-leaf stage to one plant per pot, resulting in three plants per accession (a total of 18 plants for the study). The plants were positioned on greenhouse benches fitted with trellises to which the vines were trained as they grew, keeping them separated from adjacent vines. Greenhouse temperatures were maintained at 28/26° C (day/night) and supplemental lighting used to provide a 12-hour photoperiod. Plants were watered as needed, and either a liquid fertilizer or a slow-release pellet fertilizer applied approximately every two weeks.

Pedicel lengths were recorded for four female flowers on each plant beginning on the day that the flower opened and continuing until pedicel elongation ceased. Two of these flowers were hand-pollinated on the day they opened with pollen collected from three or four male flowers of the same accession. Pedicel lengths were recorded daily until they remained unchanged for five days. Petals of the other two female flowers were secured with a small metal clip to prevent pollination (negative control), and pedicel lengths were recorded daily until ovary abscission or until pedicel lengths remained unchanged for five days. In addition to the flowers that were selected to track pedicel development, we regularly inspected the plants to search for any female flowers with especially short or long pedicels. These were also measured by using the preceding protocol. Final pedicel measurements were taken just before fruit harvest on 15 May 2009, and no changes in lengths were observed from those previously recorded when daily measurements ceased. Digital images of fruits with attached pedicels were captured by using a flatbed scanner.

Statistical tests (1) were made to compare pollinated and unpollinated flowers across accessions and the two botanical varieties for the following traits: initial pedicel length, peak pedicel length, final pedicel length, and the number of days to reach peak pedicel length. First, variances were compared by F-tests. When variances were found to be the same, means were compared with a two-sided t-test. When variances were significantly different, medians were compared with the Rank-Sum Z test for large samples.

Results and Discussion: At the time that the female flowers first opened, pedicel lengths varied between 22 and 88 mm (median = 60 mm) in the 24 flowers selected to track pedicel development in var. *anguria* and between 25 and 69 mm (median = 36.5 mm) in the 48 corresponding flowers of var. *longaculeatus*. These median values were significantly different between botanical varieties at the 0.1% level (Z=3.281). Consistent with this difference, the shortest pedicel that we observed in any plant at the time of female-flower opening was found in var. *longaculeatus* (Ames 23541) at 7 mm and the longest was observed in var. *anguria* (PI 494824) at 108 mm. For purposes of an overall species description, we calculated an overall mean female-pedicel length of 45.6 mm and a full range of 7 to 108 mm. Our shortest pedicel was as long as the longest value reported by Kirkbride (7), and our range resembled, but slightly exceeded, values reported by Howard (3) and Jeffrey (4-6).

Pollination is clearly required for full pedicel elongation. Pedicels on the 36 flowers that were not pollinated reached their peak length (median = 46.5 mm) after only 2.7 days, but pedicels on the 36 pollinated flowers grew much longer (median = 77 mm) after 5.4 days. Pedicel development over time for pollinated and unpollinated flowers displayed by botanical variety is illustrated in Figure 1. Differences in peak length and in the time to reach it were both significant at the 0.1% level (Z=-4.681, t=7.43, respectively). Final pedicel lengths (at 98% of peak pedicel lengths) were not significantly different at the 5% level from peak lengths for either pollinated or unpollinated samples. Pedicels subtending the pollinated flowers reached a peak at 1.7× of their initial lengths, and no pedicel elongated to more than 2.35× of its initial length. This is much less than the 30× elongation suggested by the values presented by Kirkbride (7).

Final pedicel lengths for pollinated samples reflect the range of variation in fruit pedicels. The 12 fruit pedicels varied between 44 and 159 mm (median = 98.5 mm) for var. *anguria* and between 32 and 122 mm (median = 77 mm) for the 24 fruits of var. *longaculeatus* (Figure 2). These median values reflected differences in initial pedicel lengths, but were not significantly different at the 5% level. The 36 fruit-pedicel lengths fell within the range reported by Jeffrey (5) of 25 to 210 mm, but only the longest pedicel exceeded the consensus range of 60 to 135 mm reported elsewhere (2, 4, 6-8). Of the two "extreme" female-flower pedicels, the shortest produced a fruit pedicel only 5 mm long (illustrated in Figure 2), while the developing fruit on the longest aborted and then the pedicel failed to elongate further.

Our findings are contrary to the description of *C. anguria* as presented in Kirkbride's monograph (7), which encompassed both botanical varieties and was used as the basis for his key. Widrlechner et al. (10) also used length of the female-flower pedicel in modifications to both Kirkbride's key (7) and to the more recent one proposed by Schaefer (9), as part of the description of *C. zambianus*.

In couplet 34 of Kirkbride's key (7), length of the female-flower pedicel was used as a secondary, supporting character to distinguish *C. anguria*. Our findings significantly reduce the separating power of this character; thus, we propose eliminating it from that couplet and from the modified couplet 34 in Widrlechner et al. (10). Schaefer (9) chose not to use length of the female-flower pedicel to distinguish *C. anguria* in his key, so no alterations are required there. However, this character should be removed from new couplet 57 in the modification of Schaefer's (9) key made by Widrlechner et al. (10). With the removal of pedicel length from this couplet, only one character, the form of the male inflorescence, would remain. We propose strengthening new couplet 57 as follows:

57a. Male inflorescences racemose; calyx lobes of male flowers narrowly triangular; pedicels of female flowers and fruits flaring from a narrow base to a wider apex..14. *C. anguria*

57b. Male inflorescences paniculate; calyx lobes of male flowers linear; pedicels of female flowers and fruits cylindrical.............................*C. zambianus*

Our findings point out some of the difficulties in observing and interpreting biological processes and phenomena from herbarium specimens. Kirkbride (7) also pointed this out in relation to the various reproductive systems present in *Cucumis*. Most species are monoecious, but deviations from monoecy can be difficult to identify from herbarium specimens unless the collector was observant and included appropriate inflorescences in specimens along with corresponding label notes. More directly related to measurements of pedicel length, the point at which anthesis occurs is also much harder to determine in herbarium specimens than it is from living material, which increases the possibility for misinterpretation when working solely from herbarium specimens.

Acknowledgements: Journal paper of the Iowa Agriculture and Home Economics Experiment Station, Ames, IA, and supported by Hatch Act and State of Iowa funds. The authors thank Rebecca Grumet, Michael Nee, and Loren Stephens for their valuable critiques.

Literature Cited

1.Cox, G. W. 1976. Laboratory Manual of General Ecology. W. C. Brown, Dubuque, IA.

2.Fernandes, R.,and A. Fernandes. 1970. 88 – Cucurbitaceae. Pages 232-289, In: A.W. Exell, A. Fernandes, and E.J. Mendes (eds.) Conspectos Florae Angolensis, Volume 4. Junta de Investigações do Ultamar & Instituto de Investigação Científica de Angola, Lisbon.

3.Howard, R. A. 1989. Flora of the Lesser Antilles: Leeward and Windward Islands. Volume 6. Dicotyledoneae - Part 3. Arnold Arboretum, Jamaica Plain, MA.

4.Jeffrey, C. 1967. Cucurbitaceae. In: E. Milne-Redhead and R. M. Polhill (eds.) Flora of Tropical East Africa. Crown Agents for Oversea Governments and Administrations, London.

5.Jeffrey, C. 1978. 86. Cucurbitaceae. Pages 414-499, In: E. Launert, J. P. M. Brenan, A. Fernandes, H. Wild (eds.) Flora Zambesiaca, Volume 4. Flora Zambesiaca Managing Committee, London.

6.Jeffrey, C. 2001. Cucurbitaceae Juss. Pages 688-717, In: W. D. Stevens, C. Ulloa Ulloa, A. Pool, and O. M. Montiel (eds.) Flora de Nicaragua. Monographs in Systematic Botany from the Missouri Botanical Garden 85(1).

7. Kirkbride, J. H., Jr. 1993. Biosystematic Monograph of the Genus *Cucumis* (Cucurbitaceae). Parkway Publishers, Boone, NC.

8.Nee, M. 1993. Cucurbitaceae. Fascicle 74. In: V. Sosa (ed.) Flora de Veracruz, Instituto de Ecologia, Xalapa, & University of California, Riverside.

9.Schaefer, H. 2007. *Cucumis* (Cucurbitaceae) must include *Cucumella, Dicoelospermum, Mukia, Myrmecosicyos,* and *Oreosyce*: A recircumscription based on nuclear and plastid DNA data. Blumea 52: 165-177.

10.Widrlechner, M. P., J. H. Kirkbride, Jr., A. G. Ghebretinsae, and K. R. Reitsma. 2008. *Cucumis zambianus* (Cucurbitaceae), a new species from northwestern Zambia. Systematic Botany 33: 732-738.

Table 1. Accessions of *Cucumis anguria* selected for measurement.

Accession Number	Taxonomy	Origin
PI 196477	*Cucumis anguria* var. *anguria*	Brazil
PI 494824	*Cucumis anguria* var. *anguria*	Zambia
PI 542135	*Cucumis anguria* var. *longaculeatus*	Botswana
Ames 22076	*Cucumis anguria* var. *longaculeatus*	Zambia
Ames 23536	*Cucumis anguria* var. *longaculeatus*	South Africa
Ames 23541	*Cucumis anguria* var. *longaculeatus*	South Africa

Fig. 1a. Pedicel development in var. *anguria*

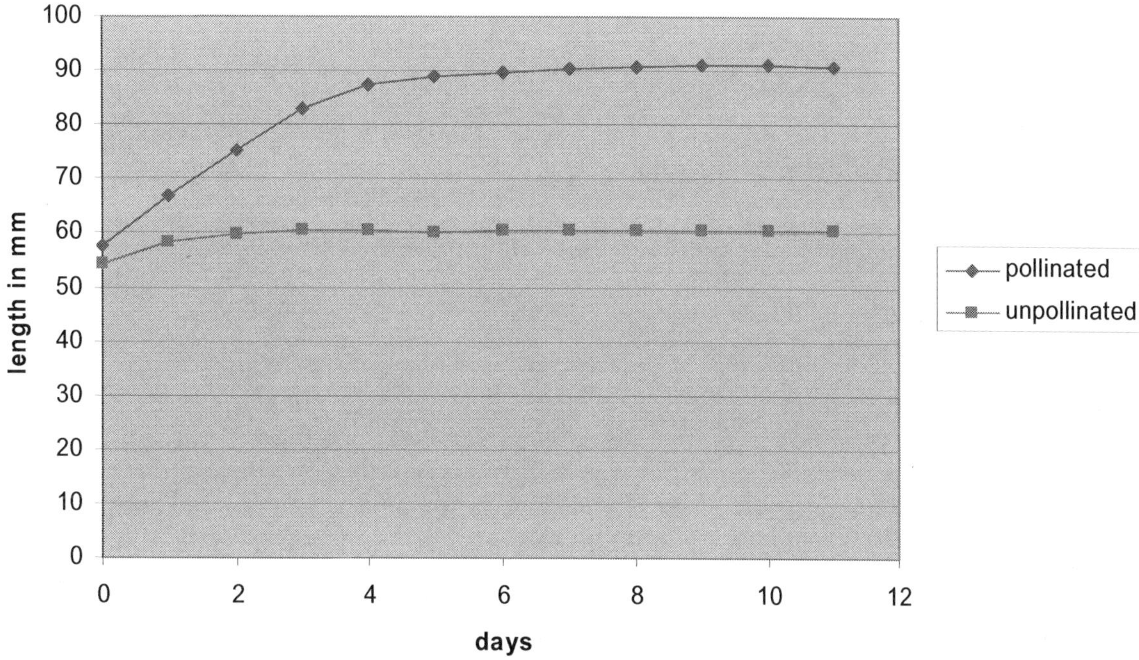

Fig. 1b. Pedicel development in var. *longaculeatus*

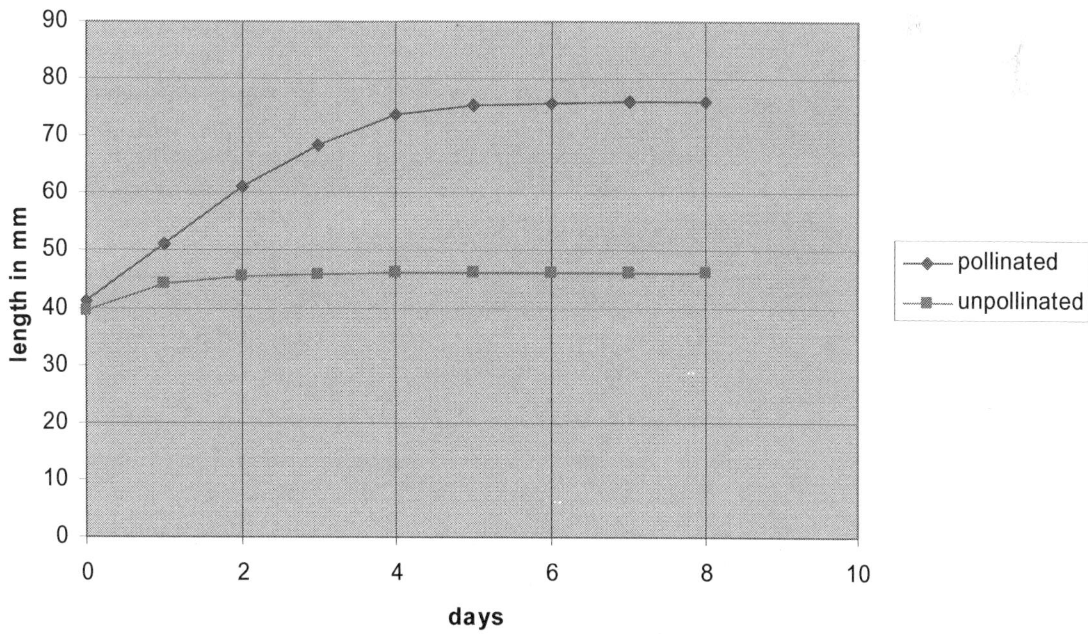

Figure 1. Mean pedicel development over time (in days after flower opening) for 12 pollinated and 12 unpollinated flowers of *C. anguria* var. *anguria* (Fig. 1a) and for 24 pollinated and 24 unpollinated flowers var. *longaculeatus* (Fig. 1b).

Figure 2. Mature fruit pedicels representing the range of length variation in *C. anguria*. From left to right, *C. anguria* var. *longaculeatus* Ames 23541 (1 fruit), PI 542135 (1 fruit), Ames 22076 (2 fruits), *C. anguria* var. *anguria* PI 196477 (2 fruits), PI 494824 (2 fruits).

Gene List for Other Genera of Cucurbitaceae 2008

Marlene Taja and Todd C. Wehner
Department of Horticultural Science, North Carolina State University, Raleigh, NC 27695-7609

Introduction

The Cucurbitaceae includes many important vegetables species, including cucumber, melon and watermelon. Those are major crop species originally from the Old World: cucumber from India; melon and watermelon from Africa (Wehner and Maynard, 2003). However, there are other important species originally from Africa such as gherkin (*Cucumis anguria*), African horned melon (*Cucumis metuliferus*), bottle gourd (*Lagenaria siceraria*); and species originally from India such as sponge gourd (*Luffa* spp.), *Melothria* (*Melothria medraspatana*) and bitter melon (*Momordica charantia*). They have fruit that are used for food, decoration, containers, utensils or sponges. The exception is *Melothria*, which has medicinal uses (Iman et al., 2006).

Gene List Update

The following list is the latest version of the gene list for the miscellaneous species and genera of the Cucurbitaceae. The genes originally were organized and summarized by Robinson (1979, 1982). This current gene list provides an update of the known genes, with 20 total mutants grouped by species.

Researchers are encouraged to send reports of new genes, as well as seed samples of lines having the gene mutant to the gene curator (Mark G. Hutton), or the assistant curator (Thomas C. Andres). Please inform us of omissions or errors in the gene list. Scientists should consult the list as well as the rules of gene nomenclature for the Cucurbitaceae (Cucurbit Gene List Committee, 1982; Robinson et al., 1976) before choosing a gene name and symbol. Please choose a gene name and symbol with the fewest characters that describes the recessive mutant, and avoid use of duplicate gene names and symbols. The rules of gene nomenclature were adopted in order to provide guidelines for naming and symbolizing genes. Scientists are urged to contact members of the gene list committee regarding rules and gene symbols. The gene curators for other genera of the Cucurbit Genetics Cooperative are collecting seeds of the type lines for use by interested researchers, and would like to receive seed samples of any of the lines listed.

This gene list has been modified from previous lists in that we have expanded the gene descriptions of the phenotypes of the gene mutants, and added genes not previously described: *Bt, S* and *P* (*Cucumis anguria*), *Prsv* (*Cucumis metuliferus*), *S* (*Lagenaria siceraria*), and *gy-1* (*Momordica charantia*).

Previous Gene Lists

- Robinson, 1979: 13 genes added, 13 genes total
- Robinson, 1982: 1 gene added, 14 genes total

West Indian Gherkin (Cucumis anguria)

Four gene loci have been described so far for West Indian gherkin. A single dominant gene produces bitter fruit: *Bt* (Koch and Costa, 1991). Another dominant gene controls resistance to Cucumber green mottle mosaic virus: *Cgm* (den Nijs, 1982). Two loci control fruit spininess: *S and P* (Koch and Costa, 1991).

African horned cucumber (*Cucumis metuliferus*)

Watermelon mosaic virus-1 resistance is controlled by a single dominant gene *Wmv* (Provvidenti and Robinson, 1972). Another single dominant gene, *Prsv* controlled resistance to Papaya ringspot virus (Provvidenti and Gonsalves, 1982). The resistant type line was PI 292190, and the susceptible type line was Acc. 2459.

Bottle Gourd (*Lagenaria siceraria*)

Red pumpkin beetle (*Aulacophora faveicollis*) resistance is controlled by a single dominant gene *Af* (Vashishta and Choudhury, 1972). Different genes affect shape and color of the fruit in bottle gourd. The genotype *bb* produces bottle-shaped fruit, and *BB* produces disk-shaped fruit. The genotype *rr* produces round fruit shape that is also recessive to the genotype *RR*, with disk-shaped fruit. The gene *db* interacts with *b* to produce an F2 of 9 club: 3 round: 4 dumbbell-shaped fruit (Tyagi, 1976). Dark green fruit color is controlled by the genotype *GG* which is dominant to the genotype *gg* with light green fruit color (Tyagi, 1976). The genotype *lb lb* controls the light brown seed coat color, but it is recessive to the genotype *Lb Lb* with brown seed coat color (Tyagi, 1976).

Four normal-leaf parents (Pusa Naveen, PBOG 13, PBOG 22 and PBOG 61) were crossed with segmented-leaf parents (PBOG 54) of bottle gourd to study the inheritance of segmented leaf shape. Normal leaf shape parents showed true breeding normal leaf type plants. However, the segmented-leaf parent (PBOG 54) surprisingly segregated in a ratio of 3 segmented: 1 normal leaf shape plants. Moreover, FIS also segregated in 1 segmented: 1 normal leaf shape suggesting that the parental cultivar PBOG 54 was heterozygous for leaf shape *gene* and the segmented leaf was dominant over normal type. The segregation in the backcrosses in to 1 segmented: 1 normal leaf type confirmed that a single dominant *gene S* is responsible for segmented leaf shape character in bottle gourd (Akhilesh and Ram, 2006).

Luffa Sponge Gourd (Smooth Luffa) (*Luffa aegyptiaca*), Luffa Ridge Gourd (Angled Luffa) (*L. acutangula*)

The gynoecious gene *g* (Choudhury and Thakur, 1965) interacts with andromonoecious gene *a* to produce the following phenotypes: monecious or trimonoecious (*AA GG*), andromonoecious (*aa GG*), gynoecious (*AA gg*), or hermaphroditic (*aa gg*) plants.

Melothria (*Melothria medraspatana*)

Small seed size (3.0 mm) is controlled by the gene *s* (Sing, 1972) that is recessive to *SS* for large seed size (3.6 mm). The gene *w* controls the white seed coat color if *ww*, if *Ww* it the color will be ashy, and black if *WW* (Sing, 1972).

Bitter Melon (*Momordica charantia*)

Light brown seed *lbs* (Ram et al., 2006) is inherited as a single gene that is recessive to dark brown. Large seed size is controlled by the gene *ls*, which is recessive to small seed size (Srivastava and Nath, 1972). White immature fruit skin is controlled by the genotype *ww* for white epicarp that is recessive to the genotype *WW* for green epicarp (Srivastava and Nath, 1972).

Ram et al. (2006) reported that gynoecism in Gy263B was controlled by a single recessive gene *gy-1*. The gynoecious plants of Gy263B had significantly longer (200 cm) vine length than their monoecious counterparts (127.5 cm).

Literature Cited

1. Akhilesh, T. and H.H. Ram. 2006. Qualitative inheritance of segmented leaf shape in bottle gourd (*Lagenaria siceraria* (Molina) Standl.). 33 (2), p. 117-121 (abstract).Vegetable Science

2.Benzioni, A., V.M. Mendlinger, S. Huyskens. 1993. Germination, fruits development, yield and post harvest characteristics of *C. metuliferus*. In: J. Janick and J.E. Simon (eds) New crops. Wiley, New York. pp. 553-557.

3.Choudhury, B. and M.R. Thakur. 1965. Inheritance of sex forms in Luffa. Indian J. Genet. Pl. Breed. 25:188-197.

4.den Nijs, A.P.M. 1982. Inheritance of resistance to cucumber green mottle virus (CGMV) in *Cucumis anguria L.* Cucurbit Genetics Coop. Rpt. 5:56-57.

5.Iman, R.A., B.L. Priya, R. Chithra, K. Shalini, V. Sharon, D. Chamundeeswari, and J. Vasantha. 2006. In vitro antiplatelet activity-guided fractionation of aerial parts of *Melothria maderaspatana*. Indian J. Pharm. Sci. 68:668-670.

6.Koch, P.S. and C.P. daCosta. 1991. Inheritance of plant and fruit characters in gherkin. Horticultura Brasileira 9 (2): 73-77 (abstract).

7.Morton, J.F. 1987. The horned cucumber alias "Kiwano" (*Cucumis metuliferus, cucurbitaceae*). Econ. Botany 41: 325-326.

8.Provvidenti, R. and D. Gonsalves. 1982. Resistance to papaya ringspot virus in *Cucumis metuliferus* and its relationship to resistance to watermelon mosaic virus 1.J. Heredity 73(3): 239-240.

9.Provvidenti, R. and R.W. Robinson. 1972. Inheritance of resistance to watermelon mosaic virus in *Cucumis metuliferus*. J. Hered. 68:56-57.

10.Ram, D., S. Kumar, M. Singh, M. Rai and G. Kalloo. 2006. Inheritance of gynoecism in bitter gourd (*Momordica charantia* L.). J. Heredity 97(3):294-295.

11.Robinson, R.W. 1979. New genes for the Cucurbitaceae. Cucurbit Genetics Coop. Rpt. 2:49-53.

12.Robinson, R.W. 1982. Update of Cucurbit gene list and nomenclature rules. Cucurbit Genetics Coop. Rpt. 5:62-66.

13.Sing, A.K. 1972. Inheritance of some seed characters in Melothria medraspatana (L.) Conz. Balwant Vidyapeeth Jour. Agr. Sci. Res. 14:56-57.

14.Srivastava, V.K. and P. Nath. 1972. Inheritance of some qualitative characters in *Momordica charantia* L. Indian J. Hort. 29:319-321.

15.Tyagi, I.D. 1976. Inheritance of some qualitative characters in bottle gourd (*Lagenaria siceraria* Standal). Indian J. Hort. 33:78-82.

16.Vashishta, R.N. and B. Choudhury. 1972. Inheritance of resistance to red pumpkin beetle in muskmelon, bottle gourd and watermelon. Proc. 3rd Intern. Symposium Sub-Trop. Hort. 1:75-81.

17.Wehner, T.C. and D.N. Maynard. 2003. Cucumbers, melons, and other cucurbits. In: S.H. Katz (editor) Encyclopedia of Food and Culture. Scribner & Sons, New York, 2014 p.

18.Wehner, T.C. and D.N. Maynard. 2003. Cucurbitaceae (vine crops). In: Encyclopedia of Life. Nature Publishing.

19. Wehner, T.C. and D.N. Maynard. 2003. Cucurbitaceae (vine crops). In: Encyclopedia of Life. Nature Publishing.

Table 1. The morphological and resistance genes of the miscellaneous genera and species of the Cucurbitaceae,

Symbol	Gene description and type lines	References
Cucumis anguria		
Bt	*Bitter fruit.* Fruit with bitter flavor due to a single dominant gene determined in the segregating populations of *Cucumis anguria x C. longipes*.	Koch and Costa, 1991
Cgm	*Cucumber green mottle resistance.*	den Nijs, 1982
S	*Spine fruit.* The fruit spininess is determined in the segregating populations of *Cucumis anguria x C. longipes*. by two pairs of independent genes.	Koch and Costa, 1991
P	*Spine fruit.* The fruit spininess is determined in the segregating populations of *Cucumis anguria x C. longipes*. by two pairs of independent genes.	Koch and Costa, 1991
Cucumis metuliferus		
Prsv	*Papaya ringspot virus resistance. Resistance to papaya ringspot virus; dominant to susceptibility.*	Provvident and Gonsalves, 1982
Wmv	*Watermelon mosaic virus resistance.* Resistance to watermelon virus-1; dominant to susceptibility.	Provvidenti and Robinson, 1972
Lagenaria siceraria		
Af	*Aulacophora foveicollis resistance.* Resistance dominant to susceptibility to the red pumpkin beetle.	Vashishta and Choudhury, 1972
b	*bottle.* Bottle-shaped fruit recessive to disk.	Tyagi, 1976
db	*dumbbell.* Interacts with b to produce F2 of 9 club: 3 round: 4 dumbbell-shaped fruit.	Tyagi, 1976
G	*Green.* Dark green fruit color; dominant to light green.	Tyagi, 1976
lb	*light brown seed.* Light brown seed coat color recessive to brown.	Tyagi, 1976
r	*round.* Round fruit; recessive to disk fruit shape.	Tyagi, 1976
S	*Segmented leaves.* A single dominant gene which is responsible for segmented leaf shape in bottle gourd from PBOG 54 (heterozygous for segmented leaf shape).	Akhilesh and Ram, 2006
***Luffa* spp.**		
g	*gynoecious.* Pistillate flowers only; interacts with a to produce monoecious or trimonoecious (*AA GG*), andromonoecious (*aa GG*), gynoecious (*AA gg*), or hermaphroditic (*aa gg*) plants.	Choudhury and Thakur, 1965
Melothria medraspatana		
s	*small seeds.* Small (3.0 mm) seed recessive to large (3.6 mm).	Sing, 1972
w	*white seeds.* White seed coat if *ww*, ashy if *Ww*, and black if *WW*.	Sing, 1972
Momordica charantia		
gy-1	*gynoecious* Recessive gene for high degree of pistillate sex expression from Gy263B (100% gynoecious line).	Ram et al., 2006
lbs	*light brown seed.* Light brown seed coat color; recessive to dark brown.	Srivastava and Nath, 1972
ls	*large seed.* Large seed size; recessive to small seed size.	Srivastava and Nath, 1972
w	*white epicarp.* White immature fruit skin; recessive to green.	Srivastava and Nath, 1972

Gene List for *Cucurbita* species, 2009

Harry S. Paris
A.R.O., Newe Ya'ar Research Center, Ramat Yishay 30-095 (Israel)

Eileen Kabelka
University of Florida, Gainesville, FL 32611 (U.S.A.)

The genus *Cucurbita* L. contains 12 or 13 species (50). As far as is known, all have a complement of 20 pairs of chromosomes (2n = 40).

This gene list for *Cucurbita* contains detailed sources of information, being modeled after the one for cucumber presented by Wehner and Staub (103) and its update by Xie and Wehner (109). In order to more easily allow confirmation of previous work and as a basis for further work, information has been included concerning the genetic background of the parents that had been used for crossing. Thus, in addition to the species involved, the cultivar-group (for *C. pepo*), market type (for *C. maxima, C. moschata*), and/or cultivar name are included in the description wherever possible.

Genes affecting phenotypic/morphological traits are listed in **Table 1**. The data upon which are based identifications and concomitant assignment of gene symbols vary considerably in their content. No attempt is made here to assess the certainty of identifications, but gene symbols have been accepted or assigned only for cases in which at least some data are presented. The genes that are protein/isozyme variants are listed in **Table 2**. It can be seen from Tables 1 and 2 that approximately 70 genes have been identified for *C. pepo* L, for *C. moschata* Duchesne 25 and for *C. maxima* Duchesne 19. For the interspecific cross of *C. maxima* × *C. ecuadorensis* Cutler & Whitaker, 29 genes have been identified, of which 25 are isozyme variants. A few genes have also been identified in four of the wild species (*C. okeechobeensis* Bailey, *C. lundelliana* Bailey, *C. foetidissima* HBK and *C. ecuadorensis*) and in several other interspecific crosses.

Some genes are listed as occurring in more than one species. This does not necessarily indicate that these genes reside at identical locations in the genome of different species.

New additions to the list of *Cucurbita* genes include a number of omissions as well as a number of new genes published after the last update. Those that had been omitted are three unnamed genes for fruit bitterness (3). They are herein designated *Bitter fruit-1, -2,* and *-3*, symbols *Bi-1, Bi-2, and Bi-3*. This has necessitated the modification of the symbols for the two previously identified genes (12, 30, 32) for *Bitter fruit* as *Bi^{max}* and *Bi-0*. Newly identified genes that have been published since the last update are: *ae* (*androecy enhancer*), *Crr-1, Crr-2,* and *Crr-3* (*Crown rot resistance-1, -2,* and *-3*), *gl-2* (*glabrous-2*), and *l-2^R* (*light type-2 Reverse striping*). The symbols *ae, Crr-1, Crr-2, and Crr-3* are herein assigned for the first time. Before choosing a gene name and symbol, researchers are urged to consult this Gene List as well as the rules of Gene Nomenclature for the Cucurbitaceae that appears near the end of this Cucurbit Genetics Cooperative Report in order to avoid confusion arising from duplication of gene names and symbols. Please contact us if you find omissions or errors in this Gene List.

Several cases of genetic linkage have been reported: *D – mo-2* (61) and *M – Wt* (*C. pepo*) (72) and *Bi – Lo-2* (*C. ecuadorensis* × *C. maxima*) (32). Some of the isozyme variants observed by Weeden & Robinson (102) were also found to be linked to one another. RAPD markers have been categorized and organized into linkage groups and are not listed here but can be found in Brown and Myers (5) and Zraidi and Lelley (111). These two maps cannot be easily compared, as they were constructed using different mapping populations; RAPD markers are population-specific. Neither map gives complete coverage of the *Cucurbita* genome. Both maps contain morphological traits, either as single genes or as quantitative trait loci (QTLs), which are listed in **Table 3**. More recently, a map for *Cucurbita pepo* has been constructed using RAPDs, AFLPs, and SSRs (27, 113). Over 300 markers were mapped, with coverage of some 2,200 cM of the genome, 20 linkage groups and a map density of 2.9 cM.

Sequenced genes can be valuable to breeders and geneticists, as the differences in the gene sequences that result in the phenotypes of interest can be used in marker-assisted selection. Unlike random markers, these gene-specific, allele-specific markers are completely linked to the genes of interest. Most of the genes sequenced in *Cucurbita* have been isolated by researchers doing comparative studies of specific genes across plant families; usually only a single allele is available. Nonetheless, we have included a list of the sequenced genes as **Table**

4 because the sequences could be useful as a starting point for breeders interested in isolating the genes from lines of differing phenotype. In addition to the genes listed here, there exists a collection of partial sequences from mRNA for genes differentially expressed during seed development in *C. pepo*. These expressed sequence tags were identified in a study of the naked seed trait. The Gene Accession numbers for these sequences are CD726806 through CD726832.

Table 1. Phenotypic/Morphologic Characteristics

Gene Symbol				
Preferred	**Synonym**	**Character**	**Species**	**Reference(s)**
a		*androecious*. Found in 'Greckie'; produces only male flowers, recessive to *A*.	*pepo*	41
*ae**		*androecy enhancer*. From cross between two vegetable-marrow cultivars, the strongly male 'Vegetable Spaghetti', *ae/ae* , and 'Bolognese', *Ae/Ae*.	*pepo*	48
B		*Bicolor*. Precocious yellow fruit pigmentation; pleiotropic, affecting fruit and foliage, modified by *Ep-1*, *Ep-2* and *Ses-B*. Originally from 'Vaughn's Pear Shaped' ornamental gourd. *B* in *C. moschata* 'Precocious PI 165561' derived from *C. pepo* through backcrossing. Complementary to *L-2* for intense orange, instead of light yellow, fruit-flesh color.	*pepo, moschata*	57, 74, 84, 91, 93
B^{max}	*B-2*	*Bicolor*. Precocious yellow fruit pigmentation, from subsp. *andreana* PI 165558	*maxima*	92, 95
Bi^{max}*	*Bi*	*Bitter* fruit. High cucurbitacin content in fruit. *Bi* from *C. maxima* subsp. *andreana* and *C. ecuadorensis*; *bi* from *C. maxima* subsp. *maxima*, including 'Queensland Blue'. Linked to *Lo-2*.	*maxima*, *maxima* × *ecuadorensis*	12, 32
Bi-0 *	*Bi*	*Bi-0* from wild Texan gourd; *bi-0* from zucchini squash. Might be identical with either *Bi-1* or *Bi-2*.	*pepo*	30
Bi-1 *		In cross of *C. pepo* × *C. argyrosperma*, three complementary dominant alleles are needed for bitterness. *Bi-1* from *C. pepo* straightneck 'Goldbar', *bi-1* from *C. argyrosperma* 'Green Striped Cushaw'.	*pepo* × *argyrosperma*	3
Bi-2 *		In cross of *C. pepo* × *C. argyrosperma*, three complementary dominant alleles are needed for bitterness. *Bi-2* from *C. pepo* straightneck 'Goldbar', *bi-2* from *C. argyrosperma* 'Green Striped Cushaw'.	*pepo* × *argyrosperma*	3
Bi-3 *		In cross of *C. pepo* × *C. argyrosperma*, three complementary dominant alleles are needed for bitterness. *Bi-3* from *C. argyrosperma* 'Green Striped Cushaw', *bi-3* from *C. pepo* straightneck 'Goldbar'.	*pepo* × *argyrosperma*	3
bl		*blue* fruit color. Incompletely recessive to *Bl* for green fruit color, in hubbard squash.	*maxima*	33
Bn		*Butternut fruit shape*, from 'New Hampshire Butternut', dominant to *bn* for crookneck fruit shape, as in 'Canada Crookneck'.	*moschata*	52
Bu	*D*	*Bush* habit. Short internodes; dominant to vine habit, *bu*, in young plant stage. In *C. pepo*, *Bu* in 'Giant Yellow Straightneck' and near-isogenic line of 'Table Queen', *bu* in 'Table Queen' acorn. In *C. maxima*, *Bu* from inbred line, *bu* from 'Delicious'. In *C. moschata*, *Bu* from inbred line, *bu* from undisclosed parent.	*pepo, maxima, moschata*	18, 31, 66, 90, 106

Gene Symbol				
Preferred	**Synonym**	**Character**	**Species**	**Reference(s)**
Cmv		*Cucumber mosaic virus resistance*, from Nigerian Local. Dominant to *cmv* for susceptiblity, from 'Waltham Butternut'.	*moschata*	4
cr		*cream* corolla. Cream to nearly white petals, *cr* from *C. okeechobeensis*; *Cr* from *C. moschata* 'Butternut' incompletely dominant (yellow petals for *Cr/cr*, and orange for *Cr/Cr*)	*moschata* × *okeechobeensis*	81
*Crr-1**		*Crown rot* resistance. Resistance to *Phytophthora capsici*, introgressed from *C. lundelliana* and *C. okeechobeensis* subsp. *okeechobeensis* into a breeding line of *C. moschata*. One of three complementary dominant genes for resistance. Genotype of the susceptible *C. moschata* 'Butterbush' is *crr-1/crr-1*.	*moschata*	56
*Crr-2**		*Crown rot* resistance. Resistance to *Phytophthora capsici*, introgressed from *C. lundelliana* and *C. okeechobeensis* subsp. *okeechobeensis* into a breeding line of *C. moschata*, One of three complementary dominant genes for resistance. Genotype of the susceptible *C. moschata* 'Butterbush' is *crr-2/crr-2*.	*moschata*	56
*Crr-3**		*Crown rot* resistance. Resistance to *Phytophthora capsici*, introgressed from *C. lundelliana* and *C. okeechobeensis* subsp. *okeechobeensis* into a breeding line of *C. moschata*. One of three complementary dominant genes for resistance. Genotype of the susceptible *C. moschata* 'Butterbush' is *crr-3/crr-3*.	*moschata*	56
cu		*cucurbitacin*-B reduced; *cu* for reduced cucurbitacin-B content of cotyledons of 'Early Golden Bush Scallop'; *Cu* for high cucurbitacin content of cotyledons of 'Black Zucchini'.	*pepo*	89
D		*Dark* stem. Series of three alleles observed in *C. pepo*: *D* for dark stem and dark intermediate-age fruit, D^s for dark stem but fruit not affected, and *d* for light stem and fruit not affected, with dominance $D > D^s > d$. *D* from 'Fordhook Zucchini', D^s from 'Early Prolific Straightneck'; *d* from 'Vegetable Spaghetti'. Epistatic to genes *l-1* and *l-2* when either is homozygous recessive; linked to *mo-2*. In *C. maxima*, only the fruit was observed: *D* for dark intermediate-age fruit from the zapallito 'La Germinadora'; *d* for light intermediate-age fruit from a variant zapallito breeding stock.	*pepo, maxima*	26, 45, 60, 61, 64, 66, 73, 86
de		*determinate* plant habit; stem lacking tendrils and terminating with female flowers. Recessive to *De* for indeterminate plant habit. *De* from 'Jeju' and 'Sokuk', *de* from inbred designated "Det".	*moschata*	42
Di		*Disc* fruit shape. From scallop squash, dominant to spherical or pyriform.	*pepo*	97, 104

Gene Symbol				
Preferred	Synonym	Character	Species	Reference(s)
Ep-1		*Extender of pigmentation-1*; modifier of *B*. *Ep-1* incompletely dominant to *ep-1* and additive with *Ep-2*. *Ep-1* from 'Small Sugar 7 × 7' pumpkin; *ep-1* from 'Table King' acorn.	*pepo*	96
Ep-2		*Extender of pigmentation-2*; modifier of *B*. *Ep-2* incompletely dominant to *ep-2* and additive with *Ep-1*. *Ep-2* from 'Table King' acorn; *ep-2* from 'Small Sugar 7 × 7' pumpkin.	*pepo*	96
Fr		*Fruit fly* (*Dacus cucurbitae*) resistance. *Fr* from 'Arka Suryamukhi', dominant to *fr* for susceptibility.	*maxima*	53
fv		*fused vein*. Fusion of primary leaf veins, subvital male gametophyte; found in hull-less-seeded pumpkin breeding line.	*pepo*	8, 9
G	*a, m*	*Gynoecious* sex expression; dominant to *g* for monoecious sex expression.	*foetidissima*	19, 24
Gb		*Green band* on inner side of base of petal, from a scallop squash; dominant to *gb*, for no band, from a straightneck squash.	*pepo*	20
gc		*green corolla*. Green, leaf-like petals, sterile; in unspecified F2 population.	*pepo*	99
*gl-1**	*gl*	*glabrous*, lacking trichomes	*maxima*	39
gl-2		*glabrous*, lacking trichomes; *gl-2* mutant found in straightneck squash	*pepo*	108
Gr	*G*	*Green rind*. Dominant to buff skin of mature fruit. *Gr* from 'Long Neapolitan', *gr* from 'Butternut'.	*moschata*	77
grl		*gray leaf*. Recessive to green leaf. Recessive *grl* derived from cross of zapallito-type line of *C. maxima* and a butternut-type line of *C. moschata*. Dominant *Grl* from zapallito-type *C. maxima*.	*maxima* × *moschata*	44
Hi		*Hard rind inhibitor*. *Hi*, for hard-rind inhibition, from *C. maxima* 'Queensland Blue'; *hi*, for no hard-rind inhibition, from *C. ecuadorensis*.	*maxima* × *ecuadorensis*	32
Hr		*Hard rind*. *Hr* for hard (lignified) rind in ornamental gourd, straightneck squash, and zucchini; *hr* for soft (non-lignified) rind in 'Small Sugar' pumpkin and 'Sweet Potato' ('Delicata'). Complementary to *Wt* for *Warty* fruit.	*pepo*	47, 85
i		*intensifier* of the *cr* gene for cream flowers. *Cr/– I/–* for intense orange or yellow flowers, *Cr/– i/i* for light orange or yellow flowers, *cr/cr I/–* for cream flowers, *cr/cr i/i* for white flowers. *I* from *C. moschata* 'Butternut', *i* from *C. okeechobeensis*.	*moschata* × *okeechobeensis*	81
I-mc	I_{mc}	*Inhibitor* of *mature* fruit color; dominant to *i-mc* for no inhibition. *I-mc* in a scallop squash.	*pepo*	10

Gene Symbol				
Preferred	**Synonym**	**Character**	**Species**	**Reference(s)**
I-T		*Inhibitor* of the *T* gene for trifluralin resistance. *I-T* from 'La Primera'; *i-t* from 'Ponca' and 'Waltham Butternut'.	*moschata*	1
l-1	*c, St*	*light fruit coloration-1*. Light intensity of fruit coloration. Series of five alleles observed in *C. pepo* which, in complementary interaction with the dominant *L-2* allele, give the following results: *L-1* for uniformly intense/dark fruit coloration, from 'Fordhook Zucchini'; $l\text{-}1^{BSt}$ for broad, contiguous intense/dark stripes, from 'Cocozelle'; $l\text{-}1^{St}$ for narrow, broken intense/dark stripes, from 'Caserta'; $l\text{-}1^{iSt}$ for irregular intense/dark stripes, from 'Beirut' vegetable marrow; *l-1* for light coloration, from 'Vegetable Spaghetti', with dominance of $L\text{-}1 > (l\text{-}1^{BSt} > l\text{-}1^{St}) \geq l\text{-}1^{iSt} > l\text{-}1$. In *C. maxima*, *L-1* from the zapallito 'La Germinadora'; *l-1* from a variant zapallito breeding stock.	*pepo, maxima*	26, 45, 62, 67, 63, 66, 67, 73, 82, 91
l-2	*r*	*light fruit coloration-2*. Light intensity of fruit coloration. Series of four alleles observed in *C. pepo*, which, in complementary interaction with dominant alleles at the *l-1* locus, give the following results: *L-2* for intense/dark fruit coloration, with *L-1* from 'Fordhook Zucchini', and intense/dark fruit stripes with $l\text{-}1^{BSt}$ from 'Cocozelle'; allele $L\text{-}2^{w}$ has delayed and weaker effect than *L-2*, from *C. pepo* subsp. *fraterna*; $l\text{-}2^{R}$ confers reversal of color, that is, stripes lighter than the background in combination with any of the striping alleles at the *l-1* locus, or completely light fruit in the presence of *L-1*, from *C. pepo* subsp. *texana* 'Delicata'; *l-2* for light coloration, from 'Vegetable Spaghetti', with dominance of $(L\text{-}2 = l\text{-}2^{R}) > L\text{-}2$. Dominant *L-2* is also complementary with *B* for intense orange, instead of light yellow, fruit-flesh color and with recessive *qi* for intense exterior color of young fruit. In *C. maxima*, *L-2* from the zapallito 'La Germinadora'; *l-2* from a variant zapallito breeding stock.	*pepo, maxima*	26, 45, 57, 65, 68, 69, 73
lo-1	*l*	*lobed leaves-1*; recessive to *Lo-1* for non-lobed leaves	*maxima*	21
Lo-2		*Lobed leaves-2*. *Lo-2* for lobed leaves in *C. ecuadorensis* dominant to *lo-2* for unlobed leaves in *C. maxima*. Linked to *Bi*.	*ecuadorensis* × *maxima*	32
lt		*leafy tendril*. Tendrils with laminae; *lt* found in ornamental gourd.	*pepo*	83
ly		*light yellow corolla*. Recessive to orange yellow; *ly* found in ornamental gourd.	*pepo*	83

Gene Symbol				
Preferred	**Synonym**	**Character**	**Species**	**Reference(s)**
M		*Mottled* leaves. *M* for silver-gray areas in axils of leaf veins, dominant to *m* for absence of silver-gray. For *C. maxima*, *M* in 'Zuni' and *m* in 'Buttercup' and 'Golden Hubbard'. For *C. pepo*, *M* in 'Caserta' and inbred of 'Striato d'Italia' cocozelle; *m* in 'Early Prolific Straightneck' and 'Early Yellow Crookneck'. For *C. moschata*, *M* in 'Hercules' and 'Golden Cushaw', *m* in butternut type. Weakly linked to *Wt*.	*pepo, maxima, moschata*	14, 72, 82, 87
Mldg		*Mottled light* and *dark green* immature fruit color; germplasm unspecified. Dominant to *mldg* for non-mottled.	*moschata*	6
mo-1		*mature orange-1*; complementary recessive gene for loss of green fruit color prior to maturity. *Mo-1* from 'Table Queen' acorn; *mo-1* from 'Vegetable Spaghetti'.	*pepo*	61
mo-2		*mature orange-2*; complementary recessive gene for loss of green fruit color prior to maturity. *Mo-2* from 'Table Queen' acorn; *mo-2* from 'Vegetable Spaghetti'. Linked to *D*.	*pepo*	61
ms-1	ms_1	*male sterile-1.* Male flowers abort before anthesis, derived from a cross involving 'Golden Hubbard', recessive to *Ms-1* for male fertile.	*maxima*	88
ms-2	ms_2	*male sterile-2.* Male flowers abort, sterility expressed as androecium shrivelling and turning brown; *ms-2* from 'Eskandarany' (PI 228241).	*pepo*	23
ms-3	*ms-2*	*male sterile-3.*	*maxima*	39
$m\text{-}zym^{mos}$		*modifier* of dominance of *zucchini yellow mosaic* virus resistance; confers resistance to otherwise susceptible Zym^{mos}/zym^{mos} heterozygotes. $M\text{-}zym^{mos}$ in 'Soler', $m\text{-}zym^{mos}$ in 'Waltham Butternut' and 'Nigerian Local'.	*moschata*	55
n	*h*	*naked* seeds. Lacking a lignified seed coat, *n* from oil-seed pumpkin.	*pepo, moschata*	29, 86, 107, 112, 113
pl		*plain light* fruit color, *pl* from 'Beirut' vegetable marrow and 'Fordhook Zucchini'; *Pl* in 'Vegetable Spaghetti'.	*pepo*	58
Pm		*Powdery mildew* resistance. Resistance to *Podosphaera xanthii*; *Pm* from *C. lundelliana*.	*lundelliana*	76
Pm-0		*Powdery mildew resistance*. Resistance to *Podosphaera xanthii*; *Pm-0* from *C. okeechobeensis* and in *C. pepo*.	*okeechobeensis, pepo*	11, 13, 37
pm-1		*powdery mildew* resistance in *C. moschata*. Series of three alleles: $pm\text{-}1^P$ for susceptibility from 'Ponca' dominant to $pm\text{-}1^L$ for resistance from 'La Primera', which is dominant to $pm\text{-}1^W$ for susceptibility in 'Waltham Butternut'.	*moschata*	2
pm-2		*powdery mildew* resistance in *C. moschata* 'Seminole', recessive to *Pm-2* for susceptibility	*moschata*	2

Gene Symbol				
Preferred	Synonym	Character	Species	Reference(s)
prv		*papaya ringspot virus resistance*, in Nigerian Local, recessive to *Prv* for susceptibility, in 'Waltham Butternut'.	*moschata*	4
qi		*quiescent intense*. Recessive to *Qi* for not intense and complementary to *L-2* for intense young fruit color; little or no effect on mature fruit. *Qi* from 'Vegetable Spaghetti'; *qi* from 'Jack O'Lantern' pumpkin and 'Verte non-coureuse d'Italie' cocozelle.	*pepo*	63, 66
Rd		*Red* skin. Red external fruit color; dominant to green, white, yellow and gray. *Rd* from 'Turk's Cap'; *rd* from 'Warted Hubbard'.	*maxima*	46
ro		*rosette* leaf. Lower lobes of leaves slightly spiraled, *ro* derived from an ornamental gourd.	*pepo*	47
s-1	*s*	*sterile.* Male flowers small, without pollen; female flower sterile. Derived from crossing 'Greengold' with 'Banana'.	*maxima*	34
s-2		*sterile.* Male flowers small, without pollen and female flower sterile; mutant in powdery mildew resistant, straightneck squash breeding line.	*pepo*	7
Ses-B		*Selective suppression* of gene *B*. Suppression in foliage of precocious yellowing conferred by *B*. *Ses-B* in straightneck breeding line dominant to *ses-B* in 'Jersey Golden Acorn'.	*pepo*	94
sl		*silverleaf* resistance. Recessive to *Sl* for susceptibility. In *C. moschata*, *Sl* from 'Soler'; *sl* from PI 162889 and butternut types. In *C. pepo*, *Sl* from 'Black Beauty' zucchini and *sl* from Zuc76 breeding line.	*moschata, pepo*	28, 110
slc		*Squash leaf curl* virus resistance; derived from *C. moschata*.	*pepo*	50
sp		*spaghetti* flesh, breaking into strands after cooking	*pepo*	49
T		*Trifluralin* resistance. Dominant to susceptibility to the herbicide; modified by *I-T*. *T* in 'La Primera'; *t* in 'Ponca' and 'Waltham Butternut'.	*moschata*	1
uml		*umbrella-like*; leaves shaped like partially opened umbrella. Recessive *uml* derived from a cross of *C. maxima* 'Warzywna' and a *C. pepo* inbred; dominant *Uml* from 'Warzywna'.	*maxima* × *pepo*	75
v		*virescent.* Yellow-green young leaves, *v* found in 'Golden Delicious'.	*maxima*	22
W		*Weak* fruit coloration. Dominant to *w* for intense-pigmented mature fruit; *W* from scallop squash. Complementary to *Wf* for white external fruit color.	*pepo*	59, 91, 97
wc		*white corolla*. Derived from 'Ispanskaya' × 'Emerald'. Recessive to *Wc* for normal orange-yellow corolla	*maxima*	40
Wf		*White flesh.* Dominant to *wf* for colored flesh. *Wf* in a scallop squash, *wf* in a straightneck squash. Complementary to *W* for white external fruit color.	*pepo*	20, 59, 97

Gene Symbol				
Preferred	Synonym	Character	Species	Reference(s)
Wmv		*Watermelon mosaic virus resistance.* From "Menina" and "Nigerian Local", dominant to *wmv* for susceptibility in 'Musquée de Provence' and 'Waltham Butternut'. May be linked with or identical to *Zym-1*.	*moschata*	4, 25
Wmv^{ecu}		*Watermelon mosaic virus* resistance. From *C. ecuadorensis*, in a cross with an unspecified *C. maxima*.	*maxima* × *ecuadorensis*	101
Wt		*Warty* fruit. Dominant to non-warted, *wt*, and complementary to *Hr*, with fruit wartiness being expressed only in the presence of the dominant *Hr* allele. *Wt* in straightneck, crookneck, and 'Delicata'; *wt* in zucchini, cocozelle, and 'Small Sugar' pumpkin. Weakly linked to *M*.	*pepo*	72, 85, 97
wyc		*white-yellow corolla*; isolated in 'Riesen-Melonen'. Recessive to *Wyc* for normal orange-yellow corolla.	*maxima*	40
Y		*Yellow* fruit color. *Y* for yellow fruit color of intermediate-age fruits, from straightneck and crookneck squash, dominant to *y* for green intermediate-age fruit color, from vegetable marrow, ornamental gourd, and cocozelle.	*pepo*	72, 82, 90, 91, 97
yg		*yellow-green* leaves and stems	*maxima*	39
Ygp		*Yellow-green placenta.* Dominant to yellow placental color. *Ygp* in a scallop squash, *ygp* in a straightneck squash.	*pepo*	20
ys		*yellow seedling*. Lacking chlorophyll; lethal	*pepo*	47
zym^{ecu}		*zucchini yellow mosaic* virus resistance, recessive to susceptibility; zym^{ecu} from *C. ecuadorensis*, Zym^{ecu} from *C. maxima* 'Buttercup'.	*ecuadorensis*	80
zym^{mos}		*zucchini yellow mosaic* virus resistance, recessive to susceptibility; zym^{mos} from 'Soler', Zym^{mos} from 'Waltham Butternut'.	*moschata*	55
Zym-0		*Zucchini yellow mosaic* virus resistance. *Zym-0* from *C. moschata* 'Nigerian Local' dominant to *zym-0* for susceptibility from 'Waltham Butternut'. Perhaps one of two separate genes for resistance in 'Nigerian Local'.	*moschata*	4, 51, 55
Zym-1		*Zucchini yellow mosaic* virus resistance. *Zym-1* from *C. moschata* 'Menina' dominant to *zym-1* for susceptibility from *C. moschata* 'Waltham Butternut'. *Zym-1* transferred via backcrossing to *C. pepo* 'True French' zucchini, in which it confers resistance through complementary interaction with *Zym-2* and *Zym-3*. *Zym-1* is either linked with *Wmv* or also confers resistance to watermelon mosaic virus.	*moschata, pepo*	25, 55, 70, 71
Zym-2		*Zucchini yellow mosaic* virus resistance-*2*. Dominant to susceptibility and complementary to *Zym-1*. *Zym-2* from *C. moschata* 'Menina'. *Zym-2* in *C. pepo* derived from *C. moschata*, in near-isogenic resistant line of 'True French' zucchini; *zym-2* from *C. pepo* 'True French'.	*moschata, pepo*	70

Gene Symbol				
Preferred	**Synonym**	**Character**	**Species**	**Reference(s)**
Zym-3		*Zucchini yellow mosaic* virus resistance-*3*. Dominant to susceptibility and complementary to *Zym-1*. *Zym-3* from *C. moschata* 'Menina'. *Zym-3* in *C. pepo* derived from *C. moschata*, in near-isogenic resistant line of 'True French' zucchini; *zym-3* from *C. pepo* 'True French'.	*moschata, pepo*	70

*Proposed new gene symbol.

Table 2. Isozyme Variants

Gene Symbol					
Preferred	Synonym	No. alleles observed	Character	Species	Reference(s)
Aat-1	*Aat*	8	*Aspartate aminotransferase-1*. Variant among accessions.	*pepo*	17, 36
Aat-3		2	*Aspartate aminotransferase-3*. Variant among wild populations.	*pepo*	17
Aat-4		3	*Aspartate aminotransferase-4*. Variant among wild populations.	*pepo*	17
Aat-mb		2	*Aspartate aminotransferase – microbody*	*maxima* × *ecuadorensis*	102
Aat-m1		2	*Aspartate aminotransferase mitochondria-1*	*maxima* × *ecuadorensis*	102
Aat-m2		2	*Aspartate aminotransferase mitochondria-2*	*maxima* × *ecuadorensis*	102
Aat-p2		2	*Aspartate aminotransferase plastid-2*	*maxima* × *ecuadorensis*	102
Acp-1		2	*Acid phosphatase-1*	*maxima* × *ecuadorensis*	102
Acp-2		2	*Acid phosphatase-2*	*maxima* × *ecuadorensis*	102
Adh		2	*Alcohol dehydrogenase*	*pepo*	105
Aldo-p		2	*Aldolase – plastid*	*maxima* × *ecuadorensis*	101
Est-1	*Est*	2	*Esterase*	*maxima* × *ecuadorensis*	100, 102
Gal-1		2	*β-galactosidase-1*	*maxima* × *ecuadorensis*	102
Gal-2		2	*β-galactosidase-2*	*maxima* × *ecuadorensis*	102
G2d-1		3	*Glycerate dehydrogenase-1*. Variant among wild populations.	*pepo*	17
G2d-2		2	*Glycerate dehydrogenase-2*. Variant among wild populations.	*pepo*	17
Got-1		5	*Glutamine oxaloacetate-1*. Variant among accessions, wild populations, and among *Cucurbita* species.	*pepo*	15, 16, 38, 105
Got-2		3	*Glutamine oxaloacetate-2*. Variant among species.	*maxima* × *ecuadorensis*	102
Gpi		2	*Glucosephosphate isomerase*. Variant among accessions.	*pepo*	36
Gpi-3		2	*Glucosephosphate isomerase-3*. Variant among wild populations.	*pepo*	17

Gene Symbol					
Preferred	**Synonym**	**No. alleles observed**	**Character**	**Species**	**Reference(s)**
Gpi-c1		2	*Glucosephosphate isomerase cytosolic-1*	*maxima* × *ecuadorensis*	102
Gpi-c2		2	*Glucosephosphate isomerase cytosolic-2*	*maxima* × *ecuadorensis*	102
Idh-1		4	*Isocitrate dehydrogenase-1*. Variant among accessions, wild populations, and *Cucurbita* species.	*pepo*	15, 16, 17, 38, 105
Idh-2		2	*Isocitrate dehydrogenase-2*. Variant among accessions, wild populations, and *Cucurbita* species.	*pepo*	15, 16, 17, 38, 105
Idh-3		2	*Isocitrate dehydrogenase-3*. Variant among accessions and populations.	*pepo*	15, 16, 17, 38
Lap-1	*Lap*	4	*Leucine aminopeptidase*. Variant among *C. pepo* accessions.	*maxima* × *ecuadorensis*; *pepo*	17, 36, 100, 102
Mdh-1	*Mdh*	7	*Malate dehydrogenase*. Variant among accessions.	*pepo*	36
Mdh-2		3	*Malate dehydrogenase-2*. Variant among accessions, wild populations, and *Cucurbita* species.	*pepo*	15, 16, 17, 38, 105
Mdh-3		3	*Malate dehydrogenase-3*. Variant among accessions, wild populations, and *Cucurbita* species.	*pepo*	15, 16, 17, 38, 105
Mdh-m1		2	*Malate dehydrogenase mitochondria-1*	*maxima* × *ecuadorensis*	102
Mdh-m2		2	*Malate dehydrogenase mitochondria-2*	*maxima* × *ecuadorensis*	102
Mdh-c2		2	*Malate dehydrogenase cytosolic-2*	*maxima* × *ecuadorensis*	102
Per-1		2	*Peroxidase-1*	*maxima* × *ecuadorensis*	102
Per-2		3	*Peroxidase-2*. Variant among accessions and wild populations.	*pepo*	15, 16, 38
Per-3		2	*Peroxidase-3*	*maxima* × *ecuadorensis*	102
Pgi-1		2	*Phosphoglucase isomerase-1*	*pepo*	15
Pgi-2		2	*Phosphoglucase isomerase-2*. Variant among *Cucurbita* species.	*pepo*	15, 38, 105
Pgi-3		4	*Phosphoglucase isomerase-3*. Variant among accessions, wild populations, and *Cucurbita* species.	*pepo*	15, 16, 38, 105
Pgm-1	*Pgm*	2	*Phosphoglucomutase*. Variant among accessions.	*pepo*	36

Gene Symbol					
Preferred	**Synonym**	**No. alleles observed**	**Character**	**Species**	**Reference(s)**
Pgm-2		4	*Phosphoglucomutase-2*. Variant among accessions, wild populations, and *Cucurbita* species.	*pepo*	15, 16, 38, 105
Pgm-5		2	*Phosphoglucomutase-5*. Variant among wild populations.	*pepo*	17
Pgm-6		2	*Phosphoglucomutase-6*. Variant among wild populations.	*pepo*	17
Pgm-c2		2	*Phosphoglucomutase cytosolic-2*	*maxima* × *ecuadorensis*	102
Pgm-p		2	*Phosphoglucomutase plastid*	*maxima* × *ecuadorensis*	102
Skd-1		6	*Shikimate dehydrogenase*. Variant among wild populations.	*pepo*	17
Skdh		5	*Shikimate dehydrogenase*. Variant among *C. pepo* accessions.	*maxima* × *ecuadorensis*; *pepo*	36, 102
Sod-1		2	*Superoxide dismutase-1*	*maxima* × *ecuadorensis*	102
Tpi-c2		2	*Triosephosphatase isomerase cytosolic-2*	*maxima* × *ecuadorensis*	102
Tpi-p2		2	*Triosephosphatase isomerase plastid-2*	*maxima* × *ecuadorensis*	102

Table 3. Mapped Phenotypic/Morphological Characteristics

Trait	Symbol	Linked Marker(s)	Recombination Distance (cM)	Reference(s)
Precocious yellow fruit	*B*	I10_1700	27.1	5
Bush growth habit	*Bu*	CMTp131	7.8	27
Dwarf	*Bu*	S1225_548, SCAR3_398	2.29	43
Leaf Mottle	*M*	H14_600 U489_1200	13.0 16.3	5
Seed Coat	*n*	AK11_340	4.4	111
Hull-less seed	*n*	CMTp58, CMTp151, CMTm115, CMTm239	1.5 - 3.6	27
Mature Fruit Color	[none given]	G17_700	9.7	5
Fruit Length	(QTL)	AE07_165, AC10_490, AJ20_420, P13_750, J01_600, AO20_1200, T08_460, AB08_540, AE09_1600		111
Fruit Width	(QTL)	AE07_165, AJ20_420, AM10_950, AG08_440		111
Fruit Length/width Ratio	(QTL)	AE07_165, AC10_490, AJ20_420, P13_750, J01_600		111
No. of Fruit Chambers	(QTL)	P13_950, AE08_470		111
Leaf Indentation	(QTL)	F10_400, K11_950, G2_400		5
Fruit Shape	(QTL)	F8_1050, B8_900, H19_500		5

Literature Cited in text and Tables 1, 2, and 3

1. Adeniji, A.A. and D.P. Coyne. 1981. Inheritance of resistance to trifluralin toxicity in *Cucurbita moschata* Poir. HortScience 16: 774–775.
2. Adeniji, A.A. and D.P. Coyne. 1983. Genetics and nature of resistance to powdery mildew in crosses of butternut with calabaza squash and 'Seminole Pumpkin'. J. Amer. Soc. Hort. Sci. 108: 360–368.
3. Borchers, E.A. and R.T. Taylor. 1988. Inheritance of fruit bitterness in a cross of *Cucurbita mixta* × *C. pepo*. HortScience 23: 603–604.
4. Brown, R.N., A. Bolanos-Herrera, J.R. Myers, and M.M. Jahn. 2003. Inheritance of resistance to four cucurbit viruses in *Cucurbita moschata*. Euphytica 129: 253–258.
5. Brown, R.N. and J.R. Myers. 2002. A genetic map of squash (*Cucurbita* sp.) with randomly amplified polymorphic DNA markers and morphological markers. J. Amer. Soc. Hort. Sci. 127: 568–575.
6. Cardosa, A.I.I., P.T. Della Vecchia, and N. Silva. 1993. Inheritance of immature fruit color in *C. moschata*. Cucurbit Genet. Coop. Rep. 16: 68–69.
7. Carle, R.B. 1997. Bisex sterility governed by a single recessive gene in *Cucurbita pepo*. Cucurbit Genet. Coop. Rep. 20: 46–47.
8. Carle, R.B. and J.B. Loy. 1996. Genetic analysis of the fused vein trait in *Cucurbita pepo* L. J. Amer. Soc. Hort. Sci. 121: 13–17.

9. Carle, R.B. and J.B. Loy. 1996. Fused vein trait in *Cucurbita pepo* L. associated with subvitality of the male gametophyte. J. Amer. Soc. Hort. Sci. 121: 18–22.
10. Clayberg, C.D. 1992. Reinterpretation of fruit color inheritance in *Cucurbita pepo* L. Cucurbit Genet. Coop. Rep. 15: 90–92.
11. Cohen, R., A. Hanan, and H.S. Paris. 2003. Single-gene resistance to powdery mildew in zucchini squash (*Cucurbita pepo*). Euphytica 130: 433–441.
12. Contardi, H.G. 1939. Estudios geneticos en *Cucurbita* y consideraciones agronomicas. Physis 18: 331–347.
13. Contin, M. 1978. Interspecific transfer of powdery mildew resistance in the genus *Cucurbita*. Ph.D. Thesis, Cornell Univ., Ithaca, New York.
14. Coyne, D.P. 1970. Inheritance of mottle-leaf in *Cucurbita moschata* Poir. HortScience 5: 226–227.
15. Decker, D.S. 1985. Numerical analysis of allozyme variation in *Cucurbita pepo*. Econ. Bot. 39: 300–309.
16. Decker, D.S. and H.D. Wilson. 1987. Allozyme variation in the *Cucurbita pepo* complex: *C. pepo* var. *ovifera* vs. *C. texana*. Syst. Bot. 12: 263–273.
17. Decker-Walters, D.S., T.W. Walters, C.W. Cowan, and B.D. Smith. 1993. Isozymic characterization of wild populations of *Cucurbita pepo*. J. Ethnobiol. 13: 55–72.
18. Denna, D.W. and H.M. Munger. 1963. Morphology of the bush and vine habits and the allelism of the bush genes in *Cucurbita maxima* and *C. pepo* squash. Proc. Amer. Soc. Hort. Sci. 82: 370–377.
19. Dossey, B.F., W.P. Bemis, and J.C. Scheerens. 1981. Genetic control of gynoecy in the buffalo gourd. J. Hered. 72: 355–356.
20. Dutta, L.P. and P. Nath. 1972. Inheritance of flower and fruit characters in squash, *Cucurbita pepo* L. 3rd Intl. Symp. Sub-Trop. Trop. Hort., pp. 69–74.
21. Dyutin, K.E. 1980. Spontaneous mutant of *Cucurbita maxima* Duch. squash with lobed leaves. Genetika 16: 176–178 (Russian).
22. Dyutin, K.E. and E.A. Afanas'eva. 1981. Inheritance of the yellow-green color of young leaves of the squash *Cucurbita maxima* Duch. Tsitologiya i Genetika 15(5): 81–82 (Russian).
23. Eisa, H.M. and H.M. Munger. 1968. Male sterility in *Cucurbita pepo*. Proc. Amer. Soc. Hort. Sci. 92: 473–479.
24. Fulks, B.K., J.C. Scheerens, and W.P. Bemis. 1979. Sex expression in *Cucurbita foetidissima* HBK. Cucurbit Genet. Coop. Rep. 2: 36.
25. Gilbert-Albertini, F., H. Lecoq, M. Pitrat, and J.L. Nicolet. 1993. Resistance of *Cucurbita moschata* to watermelon mosaic virus type 2 and its genetic relation to resistance to zucchini yellow mosaic virus. Euphytica 69: 231–237.
26. Globerson, D. 1969. The inheritance of white fruit and stem color in summer squash, *Cucurbita pepo* L. Euphytica 18: 249–255.
27. Gong, L., G. Stift, R. Kofler, M. Pachner and T. Lelley. 2008. Microsatellites for the genus *Cucurbita* and an SSR-based genetic linkage map of *Cucurbita pepo* L. Theor. Appl. Genet. 117: 37–48.
28. Gonzalez-Roman, M. and L. Wessel-Beaver. 2002. Resistance to silverleaf disorder is controlled by a single recessive gene in *Cucurbita moschata* Duchesne. Cucurbit Genet. Coop. Rep. 25: 49–50.
29. Grebenšcikov, I. 1954. Zur Vererbung der Dünnschaligkeit bei *Cucurbita pepo* L. Züchter 24: 162–166.
30. Grebenšcikov, I. 1955. Notulae cucurbitologicae II. Über *Cucurbita texana* A. Gr. und ihre Kreuzung mit einer hochgezüchteten *C. pepo*-Form. Kulturpflanze 3: 50–59.
31. Grebenšcikov, I. 1958. Notulae cucurbitologicae III. Kulturpflanze 6: 38–60.
32. Herrington, M.E. and J.P. Brown. 1988. Inheritance of leaf and fruit characteristics in *Cucurbita maxima* Duch. cv. Queensland Blue x *C. ecuadorensis* Cutler and Whitaker. Queensl. J. Agr. Anim. Sci. 45: 45–48.
33. Hutchins, A.E. 1935. The interaction of blue and green color factors in hubbard squash. Proc. Amer. Soc. Hort. Sci. 33: 514.
34. Hutchins, A.E. 1944. A male and female sterile variant in squash, *Cucurbita maxima* Duch. Proc. Amer. Soc. Hort. Sci. 44: 494–496.
35. Hutton, M.G. and R.W. Robinson. 1992. Gene list for *Cucurbita* spp. Cucurbit Genet. Coop. Rep. 15: 102–109.
36. Ignart, F. and N.F. Weeden. 1984. Allozyme variation in cultivars of *Cucurbita pepo* L. Euphytica 33: 779–785.
37. Jahn, M., H.M. Munger, and J.D. McCreight. 2002. Breeding cucurbit crops for powdery mildew resistance. In: R.R. Bélanger, W.R. Bushnell, A.J. Dik, and T.L.W. Carver, Eds., The powdery mildews, a comprehensive treatise, pp. 239–248. American Phytopathological Society, St. Paul, MN.
38. Kirkpatrick, K.J., D.S. Decker, and H.D. Wilson. 1985. Allozyme differentiation in the *Cucurbita pepo* complex: *C. pepo* var. *medullosa* vs. *C. texana*. Econ. Bot. 39: 289–299.
39. Korzeniewska, A. 1992. New genes in *Cucurbita maxima* Duch. In: R.W. Doruchowski, E. Kozik, and K. Niemirowicz-Szczytt, eds. Proc. Cucurbitaceae '92: the 5th Eucarpia Meeting on Cucurbit Genetics & Breeding, pp. 75–78.
40. Korzeniewska, A. 1996. Two independent loci for white and white-yellow corolla in *Cucurbita maxima* Duch. In: M.L. Gomez-Guillamon, C. Soria, J. Cuartero, J. Tores, and R. Fernandez-Munoz, eds. Proc. Cucurbitaceae Towards 2000: The 6th Eucarpia Meeting on Cucurbit Genetics & Breeding. Graficas Axarquia, Velez-Malaga, Spain, pp. 78–81.
41. Kubicki, B. 1970. Androecious strains of *Cucurbita pepo* L. Genet. Polon. 11: 45–51.
42. Kwack, S.N. 1995. Inheritance of determinate growth habit in *Cucurbita moschata* Poir. J. Kor. Soc. Hort. Sci. 36: 780–784.
43. Li, H., H. Zhang, G. Gong, Y. Li and C. Cui. 2006. Research of molecular markers linked to the dwarf gene in squash. In: G.J. Holmes, ed. Cucurbitaceae 2006. Universal Press, Raleigh, North Carolina, pp. 133-138.
44. Lopez-Anido, F., E. Cointry, I. Firpo, S.M. Garcia, and S. Gattuso. 2002. Inheritance of gray leaf color in a material derived from a *Cucurbita maxima* Duch. × *C. moschata* Duch. hybrid. Cucurbit Genet. Coop. Rep. 25: 46–48.
45. Lopez-Anido, F., V. Cravero, P. Asprelli, E. Cointry, I. Firpo, and S.M. Garcia. 2003. Inheritance of immature fruit color in *Cucurbita maxima* var. Zapallito (Carrière) Millan. Cucurbit Genet. Coop. Rep. 26: 48–50.
46. Lotsy, J.P. 1920. *Cucurbita* Strijdvragen. II. Eigen Onderzoekingen. Genetica 2: 1–21.
47. Mains, E.B. 1950. Inheritance in *Cucurbita pepo*. Papers Mich. Acad. Sci. Arts Letters 36: 27–30.
48. Manzano, S., V.J. Dominguez, D. Garrido, P. Gomez, and M. Jamilena. 2008. A recessive gene conferring ethylene insensitivity and androecy in *Cucurbita pepo*. In: M. Pitrat, ed. Proc. Cucurbitaceae 2008: the 9th Eucarpia Meeting on Cucurbit Genetics & Breeding, pp. 563–567.
49. Mazurek, Z. and K. Niemirowicz-Szczytt. 1992. Inheritance of spaghetti traits in *Cucurbita pepo*. In: R.W. Doruchowski, E. Kozik, and K. Niemirowicz-Szczytt, eds. Proc. Cucurbitaceae '92: the 5th Eucarpia Meeting on Cucurbit Genetics & Breeding, pp. 70–74.

50. Montes-Garcia, C.E., S. Garza-Ortega, and J.K. Brown. 1998. Inheritance of the resistance to squash leaf curl virus in *Cucurbita pepo* L. In: J.D. McCreight, ed. Cucurbitaceae '98: Evaluation and Enhancement of Cucurbit Germplasm. A.S.H.S., Alexandria, Virginia, pp. 328–330.
51. Munger, H.M. and R. Provvidenti. 1987. Inheritance of resistance to zucchini yellow mosaic virus in *Cucurbita moschata.* Cucurbit Genet. Coop. Rep. 10: 80–81.
52. Mutschler, M.A. and O.H. Pearson. 1987. The origin, inheritance, and instability of butternut squash (*Cucurbita moschata* Duchesne). HortScience 22: 535–539.
53. Nath, P., O.P. Dutta, S. Velayudhan, and K.R.M. Swamy. 1976. Inheritance of resistance to fruit fly in pumpkin. Sabrao J. 8: 117–119.
54. Nee, M. 1990. The domestication of *Cucurbita* (Cucurbitaceae). Econ. Bot. 44(3, Suppl.): 56–68.
55. Pachner, M. and T. Lelley. 2004. Different genes for resistance to zucchini yellow mosaic virus (ZYMV) in *Cucurbita moschata*. In: A. Lebeda and H.S. Paris (eds.), Proceedings of Cucurbitaceae 2004, pp. 237–243. Palacký Univ., Olomouc, Czech Republic.
56. Padley, L.D. Jr., E.A. Kabelka, and P.D. Roberts. 2009. Inheritance of resistance to crown rot caused by *Phytophthora capsici* in *Cucurbita.* HortScience 44: 211–213.
57. Paris, H.S. 1988. Complementary genes for orange fruit flesh color in *Cucurbita pepo.* HortScience 23: 601–603.
58. Paris, H.S. 1992. A recessive, hypostatic gene for plain light fruit coloration in *Cucurbita pepo.* Euphytica 60: 15–20.
59. Paris, H.S. 1995. The dominant *Wf* (*white flesh*) allele is necessary for expression of "white" mature fruit color in *Cucurbita pepo.* In: G. Lester and J. Dunlap (Eds.), Cucurbitaceae '94, pp. 219–220. Gateway, Edinburg, TX U.S.A.
60. Paris, H.S. 1996. Multiple allelism at the *D* locus in squash. J. Hered. 87: 391–395.
61. Paris, H.S. 1997. Genes for developmental fruit coloration of acorn squash. J. Hered. 88: 52–56.
62. Paris, H.S. 2000. Gene for broad, contiguous dark stripes in cocozelle squash. Euphytica 115: 191–196.
63. Paris, H.S. 2000. *Quiescent intense* (*qi*): a gene that affects young but not mature fruit color intensity in *Cucurbita pepo.* J. Hered. 91: 333–339.
64. Paris, H.S. 2000. Segregation distortion in *Cucurbita pepo.* In: N. Katzir and H.S. Paris (Eds.), Proceedings of Cucurbitaceae 2000. Acta Hort. 510: 199–202.
65. Paris, H.S. 2002. Multiple allelism at a major locus affecting fruit coloration in *Cucurbita pepo.* Euphytica 125: 149–153.
66. Paris, H.S. 2002. No segregation distortion in intersubspecific crosses in *Cucurbita pepo.* Cucurbit Genet. Coop. Rep. 25: 43–45.
67. Paris, H.S. 2003. Genetic control of irregular striping, a new phenotype in *Cucurbita pepo.* Euphytica 129: 119–126.
68. Paris, H.S. 2009. Genes for "reverse" fruit striping in squash (*Cucurbita pepo*). J. Hered. 100: 371–379.
69. Paris, H.S. and Y. Burger. 1989. Complementary genes for fruit striping in summer squash. J. Hered. 80: 490–493.
70. Paris, H.S. and S. Cohen. 2000. Oligogenic inheritance for resistance to zucchini yellow mosaic virus in *Cucurbita pepo.* Ann. Appl. Biol. 136: 209–214.
71. Paris, H.S., S. Cohen, Y. Burger, and R. Yoseph. 1988. Single gene resistance to zucchini yellow mosaic virus in *Cucurbita moschata.* Euphytica 37: 27–29.
72. Paris, H.S., A. Hanan, and F. Baumkoler. 2004. Assortment of five gene loci in *Cucurbita pepo.* In: A. Lebeda and H.S. Paris (Eds.), Proceedings of Cucurbitaceae 2004, pp. 389–394.. Palacký Univ., Olomouc, Czech Republic.
73. Paris, H.S. and H. Nerson. 1986. Genes for intense pigmentation of squash. J. Hered. 77: 403–409.
74. Paris, H.S., H. Nerson, and Y. Burger. 1985. Precocious PI 165561 and Precocious PI 165561R pumpkin breeding lines. HortScience 20: 778–779.
75. Rakoczy-Trojanowska, M. and S. Malepszy. 1999. Inheritance of umbrella-like leaf shape in materials derived form *Cucurbita maxima* ´ *C. pepo* hybrids. Cucurbit Genet. Coop. Rep. 22: 50–52.
76. Rhodes, A.M. 1964. Inheritance of powdery mildew resistance in the genus *Cucurbita.* Plant Dis. Rptr. 48: 54–55.
77. Robinson, R.W. 1987. Inheritance of fruit skin color in *Cucurbita moschata.* Cucurbit Genet. Coop. Rep. 10: 84.
78. Robinson, R.W. and M.G. Hutton. 1996. Update of gene list for *Cucurbita* spp. Cucurbit Genet. Coop. Rep. 19: 91–92.
79. Robinson, R.W. and H.S. Paris. 2000. *Cucurbita* gene list update - 2000. Cucurbit Genet. Coop. Rep. 23: 137–138.
80. Robinson, R.W., N.F. Weeden, and R. Provvidenti. 1988. Inheritance of resistance to zucchini yellow mosaic virus in the interspecific cross *Cucurbita maxima* x *C. ecuadorensis.* Cucurbit Genet. Coop. Rep. 11: 74–75.
81. Roe, N.E. and W.P. Bemis. 1977. Corolla color in *Cucurbita.* J. Hered. 68: 193–194.
82. Scarchuk, J. 1954. Fruit and leaf characters in summer squash. J. Hered. 45: 295–297.
83. Scarchuk, J. 1974. Inheritance of light yellow corolla and leafy tendrils in gourd (*Cucurbita pepo* var. *ovifera* Alef.). HortScience 9: 464.
84. Schaffer, A.A. and C.D. Boyer. 1984. The influence of gene *B* on fruit development in *Cucurbita pepo.* J. Amer. Soc. Hort. Sci. 106: 432–437.
85. Schaffer, A.A., C.D. Boyer, and H.S. Paris. 1986. Inheritance of rind lignification and warts in *Cucurbita pepo* L. and a role for phenylalanine ammonia lyase in their control. Z. Pflanzenzüchtg. 96: 147–153.
86. Schöniger, G. 1952. Vorläufige Mitteilung über das Verhalten der Testa- und Farbgene bei verschiedenen Kreuzungen innerhalb der Kürbisart *Cucurbita pepo* L. Züchter 22: 316–337.
87. Scott, D.H. and M.E. Riner. 1946. A mottled leaf character in winter squash. J. Hered. 37: 27–28.
88. Scott, D.H. and M.E. Riner. 1946. Inheritance of male sterility in winter squash. Proc. Amer. Soc. Hort. Sci. 47: 375–377.
89. Sharma, G.C. and C.V. Hall. 1971. Cucurbitacin B and total sugar inheritance in *Cucurbita pepo* related to spotted cucumber beetle feeding. J. Amer. Soc. Hort. Sci. 96: 750–754.
90. Shifriss, O. 1947. Developmental reversal of dominance in *Cucurbita pepo.* Proc. Amer. Soc. Hort. Sci. 50: 330–346.
91. Shifriss, O. 1955. Genetics and origin of the bicolor gourds. J. Hered. 46: 213–222.
92. Shifriss, O. 1966. Behavior of gene *B* in *Cucurbita.* Veg. Improv. Newsl. 8: 7–8.
93. Shifriss, O. 1981. Origin, expression, and significance of gene *B* in *Cucurbita pepo* L. J. Amer. Soc. Hort. Sci. 106: 220–232.
94. Shifriss, O. 1982. Identification of a selective suppressor gene in *Cucurbita pepo* L. HortScience 17: 637–638.
95. Shifriss, O. 1989. Relationship between the *B* genes of two *Cucurbita* species, II. Cucurbit Genet. Coop. Rep. 12: 75–78.
96. Shifriss, O. and H.S. Paris. 1981. Identification of modifier genes affecting the extent of precocious fruit pigmentation in *Cucurbita pepo* L. J. Amer. Soc. Hort. Sci. 106: 653–660.
97. Sinnott, E.W. and G.B. Durham. 1922. Inheritance in the summer squash. J. Hered. 13: 177–186.

98. Stoilova, T., N. Cholakova, and M. Markova. 2006. Variation in seed protein and isozyme patterns in *Cucurbita* cultivars. Biol. Plant. 50: 450–452.
99. Superak, T.H. 1987. A green corolla mutant in *Cucurbita pepo*. Cucurbit Genet. Coop. Rep. 10: 103.
100. Wall, J.R. and T.W. Whitaker. 1971. Genetic control of leucine aminopeptidase and esterase isozymes in the interspecific cross *Cucurbita ecuadorensis* x *C. maxima*. Biochem. Genet. 5: 223–229.
101. Weeden, N.F., R.W. Robinson, and F. Ignart. 1984. Linkage between an isozyme locus and one of the genes controlling resistance to watermelon mosaic virus 2 in *Cucurbita ecuadorensis*. Cucurbit Genet. Coop. Rep. 7: 86–87.
102. Weeden, N.F. and R.W. Robinson. 1986. Allozyme segregation ratios in the interspecific cross *Cucurbita maxima* x *C. ecuadorensis* suggest that hybrid breakdown is not caused by minor alterations in chromosome structure. Genetics 114: 593–609.
103. Wehner, T.C. and J.E. Staub. 1997. 1997 gene list for cucumber. Cucurbit Genet. Coop. Rep. 20: 66–88.
104. Whitaker, T.W. 1932. Fertile gourd-pumpkin hybrids. J. Hered. 23: 427–430.
105. Wilson, H.D. 1989. Discordant patterns of allozyme and morphological variation in Mexican *Cucurbita*. Syst. Bot. 14: 612–623.
106. Wu, T., J. Zhou, Y. Zhang, and J. Cao. 2007. Characterization and inheritance of a bush-type in tropical pumpkin (*Cucurbita moschata* Duchesne). Sci. Hort. 114: 1–4.
107. Xianglin, Z. 1987. A study on the breeding of naked kernel pumpkin and its genetic behavior. Acta Hort. Sin. 14: 115–118 (Chinese, with English summary).
108. Xiao, Q. and J.B. Loy. 2007. Inheritance and characterization of a glabrous trait in summer squash. J. Amer. Soc. Hort. Sci. 132: 327–333.
109. Xie, J. and T.C. Wehner. 2001. Gene list 2001 for cucumber. Cucurbit Genet. Coop. Rep. 24: 110–136.
110. Young, K. and E.A. Kabelka. 2009. Characterization of resistance to squash silverleaf disorder in summer squash. HortScience 44: 1213–1214.
111. Zraidi, A. and T. Lelley. 2004. Genetic map for pumpkin *Cucurbita pepo* using random amplified polymorphic DNA markers. In: A. Lebeda and H.S. Paris (Eds.), Proceedings of Cucurbitaceae 2004, pp. 507–514. Palacký Univ., Olomouc, Czech Republic.
112. Zraidi, A., M. Pachner, T. Lelley, and R. Obermayer. 2003. On the genetics and histology of the hull-less character of Styrian oil-pumpkin (*Cucurbita pepo* L.). Cucurbit Genet. Coop. Rep. 26: 57–61.
113. Zraidi, A., G. Stift, M. Pachner, A. Shojaeiyan, L. Gong, and T. Lelley. 2007. A consensus map for *Cucurbita pepo*. Mol. Breed. 20: 375–388.

Table 4. Genes with known DNA sequence

Gene Symbol*	Gene Accession	(Putative) Function	Source	Ref.
AIG-2	AY666083	aspartic protease inhibitor	*C. maxima*	**
PRB1	AY326308	phloem RNA-binding protein	*C. maxima* 'Big Max'	**
GAIP	AY32630, AY326307	gibberellic acid insensitive phloem protein (two very similar genes)	*C. maxima* 'Big Max'	**
FAD2	AY525163	omega-6 fatty acid desaturase	*C. pepo* zucchini	**
NIP1	AJ544830	Nod26-like protein	*C. pepo* zucchini	35
PP2	AY312402	phloem protein 2 lectin (includes promoter region)	*C. moschata* crookneck	**
PP2	AF150627	phloem protein 2 lectin	*C. moschata* crookneck	**
PP2	Z22647	phloem protein 2 lectin	*C. pepo* 'Autumn Gold'	61
PP2	Z17331	phloem protein 2 lectin	*C. maxima* 'Big Max'	5
PP2	L31550, L31551, L31552	phloem protein 2 (three alleles)	*C. maxima*	**
GA2OX, GA20OX, GA3OX	AJ315663, AJ302041, AJ308480, AJ302040	gibberellin oxidases (two sequences for GA2OX)	*C. maxima* 'Riesenmelone'	**
	U61385	gibberellin 20-oxidase	*C. maxima* 'Riesenmelone'	38
	U63650	gibberellin 2 beta,3 beta hydroxylase	*C. maxima* 'Riesenmelone'	39
	AJ006453	gibberellin 3 beta hydroxylase	*C. maxima* 'Riesenmelone'	**
	U61386	gibberellin dioxygenase	*C. maxima* 'Riesenmelone'	37
Moschatin 1 through 5	AF462349, AF504011, AY25646, AY27921, AY279217	ribosome-inactivating protein	*C. moschata* crookneck	**
CPS1	AB109763	copalyl diphosphate synthase; gibberellin biosynthesis	*C. maxima*	**
CPS	AF049905, AF049906	copalyl diphosphate synthase; gibberellin biosynthesis (2 genes)	*C. maxima*	55
Hsc70	AF527794, AF527795, AF527796	cell-autonomous heat shock protein; chaperonin 70 (multiple sequences)	*C. maxima*	1
	AB061204	thioredoxin h	*C. maxima*	**
Puga, Pugb, Pugc	AB055116, AB055117, AB055118	glutathione S-transferase	*C. maxima*	**
CYP88A	AF212990, AF212991	cytochrome P450; ent-kaurenoic acid oxidase (multiple alleles)	*C. maxima* 'Queensland Blue'	23

Gene Symbol*	Gene Accession	(Putative) Function	*Source*	Ref.
PP2	AF520583	phloem protein 2	*C. digitata* PI 240879	**
PP2	AF520582	phloem lectin	*C. argyrosperma* subsp. *sororia*	**
	L32700, L32701	phloem lectin	*C. argyrosperma*	5
	X56948	malate synthase	*Cucurbita* sp.*** 'Kurokawa Amakuri Nankin'	44
pMCPN60	X70867, X70868	chaperonin 60	'Kurokawa Amakuri Nankin'	59
PCPK	AY07280, AY072802	phloem calmodulin-like protein kinases	*C. maxima* 'Big Max'	66
	X55779	ascorbate oxidase	*C. maxima* 'Ebisu Nankin'	14
AAO	D55677	ascorbate oxidase	*C. maxima*	33
chitP1	AB015655	chitinase	*C. maxima* 'Ebisu Nankin'	**
PLC	AF082284	chitinase	*C. moschata* crookneck	32
PV72	AB006809	vacuolar sorting receptor	'Kurokawa Amakuri Nankin'	54
	D88420	stromal ascorbate peroxidase	'Kurokawa Amakuri Nankin'	42
	D78256	isocitrate lyase	'Kurokawa Amakuri Nankin'	41
	D70895	3-ketoacyl-CoA thiolase	'Kurokawa Amakuri Nankin'	31
	D83656	thylakoid ascorbate peroxidase	'Kurokawa Amakuri Nankin'	64
	D49433	hydroxypyruvate reductase	'Kurokawa Amakuri Nankin'	21
MP28	D45078	membrane protein	'Kurokawa Amakuri Nankin'	28
	D38132	glyoxysomal citrate synthase	'Kurokawa Amakuri Nankin'	30
	D29629	aconitase	'Kurokawa Amakuri Nankin'	19
	D16560	prepro2S albumin	'Kurokawa Amakuri Nankin'	17
	D14044	glycolate oxidase	'Kurokawa Amakuri Nankin'	58
	AF002016	acyl CoA oxidase	'Kurokawa Amakuri Nankin'	18
PP36	AF274589	cytochrome b5 reductase	*C. maxima* 'Big Max'	**
pAPX	AB070626	peroxisomal ascorbate peroxidase	'Kurokawa Amakuri Nankin'	48
CM-ACS3	AB038559	ACC synthase	*C. maxima*	62
CmATS	AB049135	acyl-(acyl-carrier protein); acyltransferase	*C. moschata* 'Shirogikuza'	**
	Y00771	glycerol-3-phosphate acyltransferase transit peptide	*C. moschata* 'Shirakikuza'	29
	AB002695	aspartic endopeptidase	*C. pepo*	24
PS-1	AF284038	phloem serpin	*C. maxima*	65
SLW	AF170086, AF170087	silverleaf whitefly-induced protein (multiple genes)	*C. pepo* zucchini 'Chefini'	60
aprX	Y17192	anionic peroxidase	*C. pepo* zucchini 'Black Beauty'	6

Gene Symbol*	Gene Accession	(Putative) Function	*Source*	Ref.
cpCPK1	U90262	calcium-dependent calmodulin-independent protein kinase	*C. pepo* zucchini	13
PP16	AF079170, AF079171	mRNA movement protein; phloem transport (multiple alleles)	*C. maxima* 'Big Max'	63
AOBP	D45066	transcription factor binding to ascorbate oxidase	*C. maxima*	34
accW	D01032	auxin-induced 1-aminocyclopropane-1-carboxylate synthase	*C. maxima* 'Ebisu'	47
	U37774	auxin-induced 1-aminocyclopropane-1-carboxylic acid synthase	*C. maxima*	46
ACC1	M58323	1-aminocyclopropane-1-carboxylate synthase	*C. pepo*	52
ACC1A, ACC1B	M61195	1-aminocyclopropane-1-carboxylate synthase (2 genes, tightly linked)	*C. pepo* zucchini	26
PHP-1	D86306	proton-translocating inorganic pyrophosphatase	*C. moschata* crookneck	**
PP1	U66277	phloem filament protein	*C. maxima* 'Big Max'	9
pfiAF4	X81647	trypsin inhibitor	*C. maxima* 'Supermarket Hybrid'	45
pfiBM7	X81447	chymotrypsin inhibitor	*C. maxima* 'Supermarket Hybrid'	45
	M15265	phytochrome	*C. pepo* zucchini 'Black Beauty'	53
NADH	M33154	nitrate reductase	*C. maxima*	11
	M36407	11S globulin beta-subunit	'Kurokawa Amakuri Nankin'	20
	AF206895	18S ribosomal RNA	*C. pepo*	**
	AF479108	26S ribosomal RNA	*C. pepo*	56
	AJ488214 EF595858 FJ915115 FJ915114 FJ915113 FJ915112 FJ915111 FJ915110 FJ915109 FJ915108 FJ915107 FJ915106 FJ915105 FJ915104 FJ915101 AM981172 AM981170 AM981169 AM981168	5.8S ribosomal RNA	*C. moschata* *C. ficifolia* *C. pepo* *C. lundelliana*	**, 7
	AY396415	5S ribosomal RNA	*C. pepo*	12
	FJ263619	16S ribosomal RNA	*C. moschata*	**

Gene Symbol*	Gene Accession	(Putative) Function	*Source*	Ref.
	DQ298735 AY357209 AY357208	18S ribosomal RNA	*C. pepo* *C. moschata*	**,4
	AF017158	25S ribosomal RNA	*C. maxima*	**
GID1b	AM745267	gibberellin receptor	*C. maxima*	**
APRX	DQ518906	class III peroxidase precursor	*C. pepo* zucchini 'Black Beauty'	10
RBP50	EU793994	polypyrimidine tract binding protein	*C. maxima* 'Big Max'	16
	AJ829947	reverse transcriptase	*C. pepo*	**
rbcL	AF206756 L21938 DQ535804 EU309692	ribulose 1,5-bisphosphate carboxylase	*C. pepo* *C. ficifolia* *C. moschata*	36,57
NACP1	FJ151402	NAC-domain containing protein	*C. maxima*	50
DNCED1	EU391616	9-cis-epoxycarotenoid dioxygenase	*C. moschata*	**
PhoH1	AB435244	alpha-1,4-glucan phosphorylase H isozyme	*C. maxima*	**
PhoL1	AB435243	alpha-1,4-glucan phosphorylase L isozyme	*C. maxima*	**
PP16-1	EU430061	16kDa phloem protein 1	*C. maxima* × *C. moschata* 'Ribenzhenmu'	**
PP16-2	EU430062	16kDa phloem protein 2	*C. maxima* × *C. moschata* 'Ribenzhenmu'	**
PP16-1	EF055181	phloem protein 1	*C. pepo*	**
PP16-2	EF055182	phloem protein 2	*C. pepo*	**
	D01033	1-aminocyclopropane-1-carboxylate synthase	*C. maxima* 'Ebisu'	27
	EF103124	mitochondrial alternative oxidase	*C. pepo*	**
matK	DQ536666 DQ536665 DQ536664	maturase K	*C. pepo* *C. digitata* *C. ficifolia*	36
trnG	EF595908	tRNA-Gly	*C. pepo*	15
	EF202177	aquaporin	*C. ficifolia*	**
	EU056338	chitinase	*C. moschata*	**
cat1	D55645	catalase	*C. pepo*	**
cat2	D55646	catalase	*C. pepo*	**
cat3	D55647	catalase	*C. pepo*	**
	AF260737	catalase	*C. pepo*	**
FTL1	EF462211 DQ865290	flowering locus T protein 1	*C. moschata* PI441726 C. maxima 'Big Max'	40

Gene Symbol*	Gene Accession	(Putative) Function	*Source*	Ref.
FTL2	DQ865291	flowering locus T protein 2	*C. maxima* 'Big Max'	40
	AB303333	glyoxalase I	*C. maxima*	**
	EF062594	Cu-Zn SOD	*C. ficifolia*	**
	EF101660 EF101661 EF101662 EF101663 EF101664 EF101665 EF101666 EF101667 EF199760 EF199759 EF199758 EF199757 EF199756 EF199755	NBS resistance protein	*C. moschata*	**
	AB002695	aspartic endopeptidase	*C. pepo*	24
DHAR	EF122791	dehydroascorbate reductase	*C. ficifolia*	**
API	DQ286449 DQ286448 DQ286447 DQ286445 DQ286444 DQ286443 DQ287856	aspartic acid proteinase inhibitor	*C. pepo* *C. maxima*	7
	EF055184 EF055183 EF055180	16 kDa phloem protein 2	*C. moschata* *C. ficifolia*	**
PP16	DQ088368 DQ088369 DQ088370 DQ088371 DQ088372 DQ088373	16 kDa. phloem protein 2	*C. maxima* 'Lefki kolokytha'	**
PATL1	DQ251455	patellin 1	*C. pepo* 'Fordhook'	49
	E02079	glycerol-3-phosphate acyltransferase	*C. moschata*	**
	AJ628045 AJ630372	histidine kinase	*C. maxima*	**
A215	X76086	14-3-3 protein endonuclease	*C. pepo*	43
EIN3	DQ023224 DQ023223	EIN3-like protein	*C. moschata*	**
aprx	Y17192	peroxidase	*C. pepo* zucchini 'Black Beauty'	**
pfiAF4	X81647	fruit trypsin inhibitor	*C. maxima* 'Supermarket Hybrid'	**

Gene Symbol*	Gene Accession	(Putative) Function	*Source*	Ref.
pfiBM7	X81447	chymotrypsin inhibitor	*C. maxima* 'Supermarket Hybrid'	**
	X73314	Gibberellin 20-oxidase	*C. maxima* 'Riesenm Elone, Gelb Genetzt'	**
	X55779	ascorbate oxidase	*Cucurbita* spp. 'Ebisu Nankin'	**
pMCPN60-2	X70867 X70868 X68606	chaperonin 60	*Cucurbita* spp. 'Kurokawa Amakuri'	**
	AJ829946 AJ829945 AJ829944	reverse transcriptase	*C. pepo*	**
NIP1	AJ544830	Nod26-like protein	*C. pepo*	35
GAIP-B	AY326307 AY326306	gibberellic acid insensitive phloem B	*C. maxima*	22
	AY663852	serine/threonine kinase-like protein	*C. ficifolia*	**
CPR	AB116239	oxidosqualene cyclase	*C. pepo*	**
CPQ	AB116238	cucurbitadienol synthase	*C. pepo*	**
	AY672635	chymotrypsin protease inhibitor	*C. maxima*	**
	AY672634	aspartic protease inhibitor	*C. maxima*	**
AIG-2	AY666083	aspartic protease inhibitor	*C. maxima*	**
AIG-1	AY666082	aspartic protease inhibitor	*C. maxima*	**
rpl2	AY396281	ribosomal protein L2	*C. pepo*	12
rpl23	AY396396	ribosomal protein L23	*C. pepo*	12
rps19	AY396376	ribosomal protein S19	*C. pepo*	12
psbC	AY396185	photosystem II protein	*C. pepo*	12
rpoB	AY396320	polymerase beta subunit	*C. pepo*	12
rps2	AY396301	ribosomal protein S2	*C. pepo*	12
FAD2	AY525163	omega-6 fatty acid desaturase	*C. pepo*	**
matR	AY453101	maturase	*C. pepo*	3
GAS1	AY379783	galactinol synthase	*C. pepo*	2
atpB	AF209573	ATP synthase beta subunit	*C. pepo*	**
Pugf	AB059484	glutathione S-transferase	*C. maxima*	25

Gene Symbol*	Gene Accession	(Putative) Function	*Source*	Ref.
nad1 nad2	AF453584 through AF453645	NADH dehydrogenase subunit 1 and 2	*C. pepo* ssp. *pepo* *C. pepo ssp. fraterna* *C. pepo ssp. ovifera* *C. pepo var. texana* *C. pepo var. ozarkana* *C. moschata* *C. maxima* *C. foetidissima* *C. argyrosperma* *C. sororia* *C. ecuadorensis* *C. andreana;* *C. okeechobeensis ssp. martinezii*	51
CmMP73	AB062669	preproMP73	*C. maxima* 'Kurokawa Amakuri Nankin'	**
CmATS1;2	AB042401 AB042400	glycerol-3-phosphate acyltransferase	*C. moschata*	**
	AF260736	glucose-6-phosphate dehydrogenase	*C. pepo*	**
	AF260735 AF260734 AF260733 AF260732	NADP-dependent malic enzyme	*C. pepo*	**
	AF260731	heat shock protein 70	*C. pepo*	**
API-2 API-1	AF038167 AF038166	aspartic proteinase inhibitor	*C. maxima*	8

* Gene symbols were assigned by the researchers isolating the gene; they have no correspondence to the official *Cucurbita* gene symbols.

**Unpublished: Genes can be submitted directly to Genbank, wthout being published in a journal.

*** 'Kurokawa Amakuri Nankin' was identified only as "*Cucurbita* sp."

Literature Cited in Table 4

1. Aoki, K., F. Kragler, B. Xoconostle-Cazares and W.J. Lucas. 2002. A subclass of plant heat shock cognate 70 chaperones carries a motif that facilitates trafficking through plasmodesmata. Proc. Natl. Acad. Sci. U.S.A. 99: 16342–16347.
2. Ayre, B.G., J.E. Blair, and R. Turgeon. 2003. Funtional and phylogenetic analyses of a conserved regulatory program in the phloem of minor veins. Plant Physiol. 133: 1229–1239.
3. Barkman, T.J., S.H. Lim, K.M. Salleh, and J. Naiz. 2004. Mitochondrial DNA sequences reveal the photosynthetic relatives of *Rafflesia*, the world's largest flower. Proc. Natl. Acad. Sci. U.S.A. 101: 787–792.
4. Bartoszewski, G., N. Katzir, and M.J. Havey. 2004. Organization of repetitive DNAs and the genomic regions carrying ribosomal RNA, cob, and atp9 genes in the cucurbit mitochondrial genomes. Theor. Appl. Genet. 108: 982–992.
5. Bostwick, D.E. and G.A. Thompson. 1993. Nucleotide sequence of a pumpkin phloem lectin cDNA. Plant Physiol. 102: 693–694.
6. Carpin, S., M. Crevecoeur, H. Greppin, and C. Penel. 1999. Molecular cloning and tissue-specific expression of an anionic peroxidase in zucchini. Plant Physiol. 120: 799–810.
7. Christeller, J.T., P.C. Farley, R.K. Marshall, A. Anandan, M.M. Wright, R.D. Newcomb, and W.A. Laing. 2006. The squash aspartic proteinase inhibitor SQAPI is widely present in the Cucurbitales, comprises a small multigene family, and is a member of the phytocystatin family. J. Mol. Evol. 63: 747–757.
8. Christeller, J.T., P.C. Farley, R.J. Ramsay, P.A. Sullivan, and W.A. Laing. 1998. Purification, characterization and cloning of an aspartic proteinase inhibitor from squash phloem exudate. Eur. J. Biochem. 254: 160–167.
9. Clark, A.M., K.R. Jacobsen, D.E. Bostwick, J.M. Dannenhoffer, M.I. Skaggs and G.A. Thompson. 1997. Molecular characterization of a phloem-specific gene encoding the filament protein, phloem protein 1 (PP1), from *Cucurbita maxima*. Plant J. 12: 49–61.
10. Cosio, C., L. Vuillemin, M. DeMeyer, C. Kevers, C. Penel, and C. Dunand. 2009. An anionic class III peroxidase from zucchini may regulate hypocotyl elongation through its auxin oxidase activity. Planta 229: 823–836.

11. Crawford, N.M., W.H. Campbell and R. Davis. 1986. Nitrate reductase from squash: cDNA cloning and nitrate regulation. Proc. Natl. Acad. Sci. U.S.A. 83: 8073–8076.
12. Decker-Walters, D.S., S.M. Chung, and J.E. Staub. 2004. Plastid sequence evolution: a new pattern of nucleotide substitutions in the Cucurbitaceae. J. Mol. Evol. 58: 606–614.
13. Ellard-Ivey, M., R.B. Hopkins, T.J White, and T.L. Lomax. 1999. Cloning, expression and N-terminal myristoylation of CpCPK1, a calcium-dependent protein kinase from zucchini (*Cucurbita pepo* L.). Plant Mol. Biol. 39: 199–208.
14. Esaka, M., T. Hattori, K. Fujisawa, S. Sakajo, and T. Asahi. 1990. Molecular cloning and nucleotide sequence of full-length cDNA for ascorbate oxidase from cultured pumpkin cells. Eur. J. Biochem. 191: 537–541.
15. Ghebretinsae, A.G., M. Thulin, and J.C. Barber. 2007. Relationships of cucumbers and melons unraveled: molecular phylogenetics of *Cucumis* and related genera (Benincaseae, Cucurbitaceae). Am. J. Bot. 94: 1256–1266.
16. Ham, B.K., J.L. Brandom, B. Xoconostle-Cazares, V. Ringgold, T.J. Lough, and W.J. Lucas. 2009. A polypryrimidine tract binding protein, pumpkin RBP50, forms the basis of a phloem-mobile ribonucleoprotein complex. Plant Cell 21: 197–215.
17. Hara-Hishimura, I., Y. Takeuchi, K. Inoue and M. Nishimura. 1993. Vesicle transport and processing of the precursor to 2S albumin in pumpkin. Plant J. 4: 793–800.
18. Hayashi, H., L. De Bellis, K. Yamaguchi, A. Kato, M. Hayashi, and M. Nishimura. 1998. Molecular characterization of a glyoxysomal long chain acyl-CoA oxidase that is synthesized as a precursor of higher molecular mass in pumpkin. J. Biol. Chem. 273: 8301–8307.
19. Hayashi, H., L. De Bellis, A. Alpi and M. Nishimura. 1995. Cytosolic aconitase participates in the glyoxylate cycle in etiolated pumpkin cotyledons. Plant Cell Physiol. 36: 669–680.
20. Hayashi, M., H. Mori, M. Nishimura, T. Akazawa and I. Hara-Nishimura. 1988. Nucleotide sequence of cloned cDNA coding for pumpkin 11-S globulin beta subunit. Eur. J. Biochem. 172: 627–632.
21. Hayashi, M., R. Tsugeki, M. Kondo, H. Mori and M. Nishimura. 1996. Pumpkin hydroxypyruvate reductases with and without a putative C-terminal signal for targeting to microbodies may be produced by alternative splicing. Plant Mol. Biol. 30: 183–189.
22. Haywood, V., T.S. Yu, N.C. Huang, and W.J. Lucas. 2005. Phloem long-distance trafficking of Gibberellic acid-insensitive RNA regulates leaf development. Plant J. 42: 49–68.
23. Helliwell, C.A., M.R. Olive, L. Gebbie, R. Forster, W.J. Peacock, and E.S. Dennis. 2000. Isolation of an ent-kaurene oxidase cDNA from *Cucurbita maxima*. Aust. J. Plant Physiol. 27: 1141–1149.
24. Hossain, M.Z. and M. Fujita. 2002. Purification of a phi-type glutathione S-transferase from pumpkin flowers, and molecular cloning of its cDNA. Biosci. Biotechnol. Biochem. 66: 2068–2076.
25. Hiraiwa, N., M. Kondo, M. Nishimura, and I. Hara-Nishimura. 1997. An aspartic endopeptidase is involved in the breakdown of propeptides of storage proteins in protein-storage vacuoles of plants. Eur. J. Biochem. 246: 133–141.
26. Huang, P.L., J.E. Parks, W.H. Rottmann, and A. Theologis. 1991. Two genes encoding 1-aminocyclopropane-1-carboxylate synthase in zucchini (*Cucurbita pepo*) are clustered and similar but differentially regulated. Proc. Natl. Acad. Sci. U.S.A. 88: 7021–7025.
27. Imaseki, H., K. Yamazaki, H. Mori, and N. Nakagawa. 1991. Cloning of a complementary DNA for auxin-induced 1-aminocyclopropane-1-carboxylase synthase and differential expression of the gene by auxin and wounding. Plant Cell Physiol. 32: 1153–1163.
28. Inoue, K., Y. Takeuchi, M. Nishimura and I. Hara-Nishimura. 1995. Characterization of two integral membrane proteins located in the protein bodies of pumpkin seeds. Plant Mol. Biol. 28: 1089–1101.
29. Ishizaki, O., I. Nishida, K. Agata, G. Eguchi and N. Murata. 1988. Cloning and nucleotide sequence of cDNA for the plastid glycerol-3-phosphate acyltransferase from squash. FEBS Lett. 238: 424–430.
30. Kato, A., M. Hayashi, H. Mori, and M. Nishimura. 1995. Molecular characterization of a glyoxysomal citrate synthase that is synthesized as a precursor of higher molecular mass in pumpkin. Plant Mol. Biol. 27: 377–390.
31. Kato, A., M. Hayashi, Y. Takeuchi and M. Nishimura. 1996. cDNA cloning and expression of a gene for 3-ketoacyl-CoA thiolase in pumpkin cotyledons. Plant Mol. Biol. 31: 843–852.
32. Kim, M.G., K.O Lee, N.E. Cheong, Y.O. Choi, J.H. Jeong, M.J. Cho, S.C. Kim, and S.Y. Lee. 1999. Molecular cloning and characterization of a class III chitinase in pumpkin leaves, which strongly binds to regenerated chitin affinity gel. Plant Sci. 147: 157–163.
33. Kisu, Y., Y. Harada, M. Goto and M. Esaka. 1997.Cloning of the pumpkin ascorbate oxidase gene and analysis of a cis-acting region involved in induction by auxin. Plant Cell Physiol. 38: 631–637.
34. Kisu, Y., T. Ono, N. Shimofurutani, M. Suzuki, and M. Esaka. 1998. Characterization and expression of a new class of zinc finger protein that binds to silencer region of ascorbate oxidase gene. Plant Cell Physiol. 39: 1054–1064.
35. Klebl, F., M. Wolf, and N. Sauer. 2003. A defect in the yeast plasma membrane urea transporter Dur3p is complemented by CpNIP1, a Nod26-like protein from zucchini (*Cucurbita pepo* L.), and by *Arabidopsis thaliana* delta-TIP or gamma-TIP. FEBS Lett. 547: 69–74.
36. Kocyan, A., L.B. Zhang, H. Schaefer, and S.S. Renner. 2007. A multi-locus chloroplast phylogeny for the Cucurbitaceae and its implications for character evolution and classification. Mol Phylogenet. Evol. 44: 553–577.
37. Lange, T. 1997. Cloning gibberellin dioxygenase genes from pumpkin endosperm by heterologous expression of enzyme activities in *Escherichia coli*. Proc. Natl. Acad. Sci. U.S.A. 94: 6553–6558.
38. Lange, T., P. Hedden, and J.E. Graebe. 1994. Expression cloning of a gibberellin 20-oxidase, a multifunctional enzyme involved in gibberellin biosynthesis. Proc. Natl. Acad. Sci. U.S.A. 91: 8552–8556.
39. Lange, T., S. Robatzek, and A. Frisse. 1997. Cloning and expression of a gibberellin 2 beta,3 beta-hydroxylase cDNA from pumpkin endosperm. Plant Cell 9: 1459–1467.
40. Lin, M.K., H. Belanger, Y.J. Lee, E. Varkonyl-Gasic, K. Taoka, E. Miura, B. Xoconostle-Cazares, K. Gendler, R.A. Jorgensen, B. Phinney, T.J. Lough, and W.J. Lucas. 2007. Flowering locus T protein may act as the long-distance florigenic signal in the cucurbits. Plant Cell 19: 1488–1506.
41. Mano, S., M. Hayashi, M. Kondo, and M. Nishimura. 1996. cDNA cloning and expression of a gene for isocitrate lyase in pumpkin cotyledons. Plant Cell Physiol. 37: 941–948.
42. Mano, S., K. Yamaguchi, M. Hayashi, and M. Nishimura. 1997. Stromal and thylakoid-bound ascorbate peroxidases are produced by alternative splicing in pumpkin. FEBS Lett. 413: 21–26.

43. Markiewicz, E., R. Rzepecki, and J. Szopa. 1994. Molecular cloning and sequencing of the cDNA encoding plant nuclear matrix endonuclease. Acta Biochem. Pol. 41: 137–138.
44. Mori, H., Y. Takeda-Yoshikawa, I. Hara-Nishimura, and M. Nishimura. 1991. Pumpkin malate synthase. Cloning and sequencing of the cDNA and northern blot analysis. Eur. J. Biochem. 197: 331–336.
45. Murray, C. and J.T. Christeller. 1995. Purification of a trypsin inhibitor (PFTI) from pumpkin fruit phloem exudate and isolation of putative trypsin and chymotrypsin inhibitor cDNA clones. Biol. Chem. Hoppe-Seyler 376: 281–287.
46. Nakagawa, N., Y. Kamiya, and H. Imaseki. 1995. Nucleotide sequence of an auxin-regulated 1-aminocyclopropane-1-carboxylic acid synthase gene (Accession No. U37774) from *Cucurbita maxima* Duch. (PGR95-110). Plant Physiol. 109: 1499.
47. Nakajima, N., H. Mori, K. Yamazaki, and H. Imaseki. 1990. Molecular cloning and sequence of a complementary DNA encoding 1-aminocyclopropane-1-carboxylate synthase induced by tissue wounding. Plant Cell Physiol. 31: 1021–1029.
48. Nito, K., K. Yamaguchi, M. Kondo, M. Hayashi, and M. Nishimura. 2001. Pumpkin peroxisomal ascorbate peroxidase is localized on peroxisomal membranes and unknown membranous structures. Plant Cell Physiol. 42: 20–27.
49. Peterman, T.K., A.S. Sequeira, J.A. Samia, and E.E. Lunde. 2006. Molecular cloning and characterization of patellin1, a novel sec14-related protein, from zucchini (*Cucurbita pepo*). J. Plant Physiol. 163: 1150–1158.
50. Ruiz-Medrano, R., B. Xoconostle-Cazares, and W.J. Lucas. 1999. Phloem long-distance transport of CmNACP mRNA: implications for supracellular regulation in plants. Development 126: 4405–4419.
51. Sanjur, O.I., D.R. Piperno, T.C. Andres, and L. Wessel-Beaver. 2002. Phylogenetic relationships among domesticated and wild species of *Cucurbita* (Cucurbitaceae) inferred from a mitochondrial gene: implications for crop plant evolution and areas of origin. Proc. Natl. Acad. Sci. U.S.A. 99: 535–540.
52. Sato, T., P.W. Oeller, and A. Theologis. 1991. The 1-aminocyclopropane-1-carboxylate synthase of *Cucurbita*. Purification, properties, expression in *Escherichia coli*, and primary structure determination by DNA sequence analysis. J. Biol. Chem. 266: 3752–3759.
53. Sharrock, R.A., J.L. Lissemore, and P.H. Quail. 1986. Nucleotide and amino acid sequence of a *Cucurbita* phytochrome cDNA clone: identification of conserved features by comparison with *Avena* phytochrome. Gene 47: 287–295.
54. Shimada, T., M. Kuroyanagi, M. Nishimura, and I. Hara-Nishimura. 1997. A pumpkin 72-kDa membrane protein of precursor-accumulating vesicles has characteristics of a vacuolar sorting receptor. Plant Cell Physiol. 38: 1414–1420.
55. Smith, M.W., S. Yamaguchi, T. Ait-Ali, and Y. Kamiya. 1998. The first step of gibberellin biosynthesis in pumpkin is catalyzed by at least two copalyl diphosphate synthases encoded by differentially regulated genes. Plant Physiol. 118: 1411–1419.
56. Soltis, D.E., A.E. Senters, M.J. Zanis, S. Kim, J.D. Thompson, P.S. Soltis, L.P. Ronse de Craene, P.K. Endress, and J.S. Farris. 2003. Gunnerales are sister to other core eudicots: implications for the evolution of pentamery. Am. J. Bot. 90: 461–470.
57. Swensen, S.M., B.C. Mullin, and M.W. Chase. 1994. Phylogenetic affinities of Datiscaceae based on an analysis of nucleotide sequences from the plastid rbcL gene. Syst. Bot. 19: 157–168.
58. Tsugeki, R., I. Hara-Nishimura, H. Mori, and M. Nishimura. 1993. Cloning and sequencing of cDNA for glycolate oxidase from pumpkin cotyledons and northern blot analysis. Plant Cell Physiol. 34: 51–57.
59. Tsugeki, R., H. Mori, and M. Nishimura. 1992. Purification, cDNA cloning and Northern-blot analysis of mitochondrial chaperonin 60 from pumpkin cotyledons. Eur. J. Biochem. 209: 453–458.
60. van de Ven, W.T., C.S. LeVesque, T.M. Perring, and L.L. Walling. 2000. Local and systemic changes in squash gene expression in response to silverleaf whitefly feeding. Plant Cell 12: 1409–1423.
61. Wang, M.B., D. Boulter, and J.A. Gatehouse. 1994. Characterization and sequencing of cDNA clone encoding the phloem protein PP2 of *Cucurbita pepo*. Plant Mol. Biol. 24: 159–170.
62. Watanabe, T., H. Fujita, and S. Sakai. 2001. Effects of jasmonic acid and ethylene on the expression of three genes for wound-inducible 1-aminocyclopropane-1-carboxylate synthase in winter squash (*Cucurbita maxima*). Plant Sci. 161: 67–75.
63. Xoconostle-Cazares, B., Y. Xiang, R. Ruiz-Medrano, H.L. Wang, J. Monzer, B.C. Yoo, K.C. McFarland, V.R. Franceschi, and W.J. Lucas. 1999. Plant paralog to viral movement protein that potentiates transport of mRNA into the phloem. Science 283: 94–98.
64. Yamaguchi, K., M. Hayashi, and M. Nishimura. 1996. cDNA cloning of thylakoid-bound ascorbate peroxidase in pumpkin and its characterization. Plant Cell Physiol. 37: 405–409.
65. Yoo, B.C., K. Aoki, Y. Xiang, L.R. Campbell, R.J. Hull, B. Xoconostle-Cazares, J. Monzer, J.Y. Lee, D.E. Ullman, and W.J. Lucas. 2000. Characterization of *Cucurbita maxima* phloem serpin-1 (CmPS-1). A developmentally regulated elastase inhibitor. J. Biol. Chem. 275: 35122–35128.
66. Yoo, B.C., J.Y. Lee, and W.J. Lucas. 2002. Analysis of the complexity of protein kinases within the phloem sieve tube system. Characterization of *Cucurbita maxima* calmodulin-like domain protein kinase 1. J. Biol. Chem. 277: 15325–15332.

2008-2009 CGC Membership Directory

Aboul-Nasr, M. Hossam. Dept. Horticulture; Faculty of Agriculure; Assiut University; Assiut, Egypt.
email: aboul-nasr@mailcity.com. Vegetable production and plant tissue culture.

Abad Martin, Jesus. Ctra. de Málaga 34, Bajo; 04710 - Sta. María del Aguila; Almería, Spain.
email: jabad@setaseeds.com. Cucumber and melon breeding.

Andres, Thomas C. The Cucurbit Network; 5440 Netherlands Ave. #D24; Bronx, NY 10471- 2321, USA.
email: tom@cucurbit.org. *Cucurbita* systematics.

Attard, Everaldo. University of Malta; Institute of Agriculture; Msida, MSD2080, Malta.
email: everaldo.attard@um.edu.mt. Research on the economic importance of *Ecballium elaterium* (squirting cucumber).

Aurangabadkar, Laxman. Ankur Seeds Pvt. Ltd.; 27, New Cotton Market Layout; Nagpur, Maharashtra 440018, India.
email: makrand@nagpur.dot.net.in. Genetics of gynoecious trait; bisexual & wild spp.; mildews & virus; sex expression in ridgegourd.

Bal, Eric. Burg. Crezeelaan; 2678 ZG, De Lier, Netherlands.
email: e.bal@rijkzwaan.nl. Cucumber; melon.

Bao, HaiQing. Xinjiang Western China Seed Group; No. 25 Luzhou Road; Changji, Xinjiang 831100, P.R. China.
email: baohaiq@hotmail.com; bhqxj@163.com. Watermelon & melon breeding; hybrid seed production techniques; variety evaluation.

Baudracco-Arnas, Sylvie. A.S.L.; Site Agroparc, 755 Chemin des Meinajaries; Bâtiment Orion, B.P. 11202; 84911, Avignon Cedex 9, France.
email: sba.asl@wanadoo.fr. Melon molecular biology.

Behera, Tusar. Division of Vegetable Science; Indian Agric. Research Institute; New Delhi, 110012, India.
email: tusar@rediffmail.com.

Bell, Duane. 17919 County Rd. B; Wauseon, OH 43567-9458, USA.
email: duaneb@ruppseeds.com.

Belotserkovsky, Harel. Hazera Genetics; Mivhor; MP Lachish Darom; 79354, Kiryat Gat, Israel.
email: harelb@limagrain.com.

Berenji, Janos. Inst. Field & Vegetable Crops; 21470, Backi Petrovac, Serbia.
email: berenji@eunet.rs.

Beronilla, Renita. East-West Seeds; Km 54 Cagayan Valley Rd.; Sampaloc, San Rafael 3008; Bulacan, Philippines.
email: Renita.beronilla@eastwestseed.com. Melon, watermelon, squash breeding.

Bertrand, Francois. Seminis Vegetable Seeds France; Mas de Rouzel; Chemin des Canaux, CS 17270; 30918, Nimes Cedex 2, France.
email: francois.bertrand@seminis.com.

Block, Charles C. North Central Regional Plant Introduction; G212 Agronomy Bldg.; Iowa State Univ.; Ames, Iowa 50011-0070, USA.
email: charles.block@ars.usda.gov.

Boissot, Nathalie. INRA GAFL; Domaine St Maurice; BP 94; 84143, Montfavet Cedex, France.
email: boissot@avignon.inra.fr.

Boyhan, George E. UGA, Southeast District Coop. Extention; P.O. Box 8112, GSU; Statesboro, GA 30460-8112, USA.
email: gboyhan@uga.edu. Pumpkin and watermelon breeding.

Brown, Rebecca. Dept. of Plant Sciences; University of Rhode Island; Woodward Hall, Alumni Dr.; Kingston, RI 02881, USA.
email: brownreb@uri.edu.

Buil Benedi, María Angeles. CLAUSE Spain, S.A.; Centro de Investigación; Apartado de Correos, 17; 04745, La Mojonera Almería, Spain.
email: maria-angeles.buil@clause-vegseeds.com.

Burger, Yosi. ARO, New Ya'ar Research Center; P.O. Box 1021; Ramat Yishay, 30095, Israel.
email: burgery@volcani.agri.gov.il.

Chen, Jin Feng. Nanjing Agricultural University; Dept. of Horticulture; Nanjing, 210095, China.
email: jfchen@njau.edu.cn. Cucumber breeding.

Cho, Myeong-Cheoul. National Horticultural Research Inst. RDA; #475 Imok-Dong; Jangan-gu, Suwon, 440-706, Korea.
email: chomc@rda.go.kr. Breeding disease resistant squash varieties and pepper breeding.

Chuanchai, Vinich. Chiatai Seeds Co., Ltd.; 299-301 Songsawad Road; Samphantawong District; Bangkok, 10900, Thailand.
email: chiatai@ksc.th.com.

Cohen, Roni. ARO, Newe Ya'ar Research Center; P.O. Box 1021; Ramat Yishay; 30095 Israel
email: ronico@volcani.agri.gov.il. Plant pathology.

Connolly, Bryan. 87 Bassets Bridge Rd.; Mansfield Center, CT 06250, USA.
email: bryan.connolly@uconn.edu; connollybryan@hotmail.com. Squash, melons, watermelon growing & breeding.

Crosby, Kevin. Texas A&M University; 2415 East Hwy 83; Weslaco, TX 78596, USA.
email: k-crosby@tamu.edu. Myrothecium stem canker on melon.

Dane, Fenny. Auburn University; 101 Funchess Hall; Auburn, AL 36849, USA.
email: danefen@auburn.edu. *Citrullus* genomics.

Davidi, Haim. Hazait 3; Moshav, Beit Elazari 76803, Israel.
email: haimdavi@zahav.net.il.

Davis, Angela. USDA, ARS South Central Agric. Res. Lab; P.O. Box 159; Hwy. 3 West; Lane, OK 74555, USA.
email: angela.davis@lane-ag.org. Germplasm improvement.

Dawson, Halina. Content Manager for Environmental Sciences; CABI, Head Office; Nosworthy Way; Wallingford, Oxfordshire OX10 8DE, United Kingdom.
email: h.dawson@cabi.org.

de Groot, Erik. Nunhems Italy SRL; Via Ghiarone, 2; 40019, Sant' Agata Bolognese BO, Italy.
email: Erik.degroot@nunhems.com. Watermelon breeding.

de Hoop, Simon Jan. East-West Seed Co.; 50/1 Moo 2; Sainoi-Bang Bua Thong Road; Sainoi, Nonthaburi 11150, Thailand.
email: simon.dehoop@eastwestseed.com. Cucurbit breeding.

De Ruiter, Wouter. De Ruiter Seeds; Leeuwenhoekweg 52; 2661 CZ, Bergschenhoek , Netherlands.
email: carla.schoonus@deruiterseeds.com. Cucurbit breeding.

Deleu, Wim. Ramiro Arnedo Semillas; Paraje La Molina, 54; 04716, Las Norias de Daza Almería, Spain.
email: wd@ramiroarnedo.com.

Den Hertog, Maarten. Riyk Zwann b.v.; 04720 Aguadulce, Almeria, Spain.
email: Maarten.den.hertog@rijkzwaan.nl.

Dombrowski, Cory. 137 E. Lake Dr.; Lehigh Acres, FL 33936, USA.
email: cdombrowski@sakata.com. Cucurbit variety evaluation and melon breeding.

Duangsong, Usa. 26/1 Moo 4 Bomnakok; Tambon Anghin; Amphor Pakthio; Ratchaburi Provence, 70140, Thailand.
email: usaduang@loxinfo.co.th; usaduang@loxinfo.co.th.

Everts, Kathryne. 27664 Nanticoke Rd; Salisbury, MD 21801, USA.
email: keverts@umd.edu. Fusarium wilt on watermelon; Gummy stem blight on watermelon and other cucurbits; Cover crops for managing cucurbit diseases (Fusarium wilt on watermelon and fruit rot on pumpkin).

Ficcadenti, Nadia. CRA-ORA; Unita di Ricera per l'Orticoltora; Via Salaria 1; 63030, Monsampolo del Tronto (A.P.), Italy.
email: nadiaf@insinet.it.

Frobish, Mark. 809 W. Delaware; Urbana, IL 61801, USA.
email: mfrobish@abbottcobb.com. Squash; sweetcorn.

Froese, David C.. 32971 CR 34;La Junta, CO 81050, USA.
email: david@coloradoseeds.com.

Furuki, Toshi. Manager of Breeding Dept. 1; Kakegawa Research Center; Sakata Seed Corporation; 1743-2 Yoshioka, Kakegawa, 436-0115, Japan.
email: t.furuki@sakata-seed.co.jp.

Gabor, Brad. Seminis Vegetable Seeds; 37437 State Hwy 16; Woodland, CA 95695, USA.
email: brad.gabor@seminis.com. Plant pathology.

Gardingo, Jose Raulindo. 142 Albert Einstein; Uvaranas; Ponta Grossa, PR 84032-015, Brazil.
email: jrgardin@uepg.br.

Garza-Ortega, Sergio. Plan de Iguala 66; Col. Mision del Sol; 83100, Hermosillo Sonora, Mexico.
email: sgarzao@prodigy.net.mx. Breeding of *Cucurbita* spp.; testing new muskmelon lines.

Gatto, Gianni. Esasem Spa; Via G. Marconi 56; 37052, Casaleone (VR), Italy.
email: ggatto@esasem.com; gpadox@gmail.com.

Goldman, Amy P.. 164 Mountain View Road; Rhinebeck, NY 12572, USA.
email: agoldthum@aol.com. Heirloom melons and watermelons; ornamental gourds; garden writing.

Gómez-Guillamón, Maria L. Estación Experimental 'La Mayora' CSIC; 29750 Algarrobo, Malaga, Spain.
email: guillamon@eelm.csic.es.

Grant, Doug. 326 c Patumahoe Road; RD 3; Pukekohe, 2678 , New Zealand.
email: doug.grant@xtra.co.nz. Breeding & genetics of *Cucurbita maxima* and *moschata.*

Groff, David. 530 Mt. Olive Church Rd.; Tifton, GA 31794, USA.
email: dave_groff@yahoo.com. Breeding *C. pepo* gourds, *C. maxima* pumpkins and cucumbers.

Grumet, Rebecca. Dept. of Horticulture; Graduate Program in Genetics; Michigan State University; East Lansing, MI 48824-1325, USA.
email: grumet@msu.edu. Disease resistance, gene flow, tissue culture and genetic engineering.

Guner, Nihat. Sakata Seed America; P.O. Box 1118; Lehigh Acres, FL 33970-1118, USA.
email: nguner@sakata.com. Watermelon breeding.

Gusmini, Gabriele. Syngenta Seeds; 10290 Greenway Rd.; Naples, FL 34114, USA.
email: gabe.gusmini@syngenta.com. Squash breeding.

Hagihara, Toshitsugu. Hagihara Farm Co., Ltd.; 984 Hokiji; Tawaramoto, Shiki Nara 636-0220, Japan.
email: cucurbit@mahoroba.ne.jp.

Haizhen, Li. Beijing Vegetable Research Center; P.O. Box 2443; Beijing, 100097, P.R. China.
email: lihaizhen@nercv.com. *Cucurbita* sp.

Havey, Michael J. USDA/ARS, Dept. of Horticulture; University of Wisconsin; 1575 Linden Dr.; Madison, WI 53706, USA.
email: mjhavey@wisc.edu.

Herrington, Mark. Maroochy Research Station; Dept Primary Industries & Fisheries; P.O. Box 5083, SCMC; Nambour, Queensland 4560, Australia.
email: mark.herrington@dpi.qld.gov.au. *Cucurbita* breeding.

Hertogh, Kees. Nickerson-Zwaan BV; P.O. Box 28; 4920 AA, Made, Netherlands.
email: kees.hertogh@nickerson-zwaan.com.

Himmel, Phyllis. Seminis Vegetable Seeds; 37437 State Highway 16; Woodland, CA 95695, USA.
email: phyllis.himmel@seminis.com. Director of Pathology and Viral disease of Cucurbits.

Hofstede, Rene. Keygene N.V.; P.O. Box 216; 6700AE, Wageningen, Netherlands.
email: rene.hofstede@keygene.com. Molecular genetic research in all Cucurbitaceae.

Holman, Bohuslav. 1420 Bzinska Str.; 69681, Bzenec, Czech Republic.
email: bholman@iol.cz. Cucumber breeding and seed production.

Hoogland, Jan. Bejo Zaden BV; P.O. Box 50; 1749, ZH Warmenhuizen, Netherlands.
email: j.hoogland@bejo.nl.

Huan, Jin. Syngenta Seeds Co.; 21435 Road 98; Woodland, CA 95695-9399, USA.
email: Jin.huan@syngenta.com. Watermelon breeding.

Ignart, Frederic. Centre de Recherche CLAUSE TEZIER; Domaine de Maninet; Route de Beaumont; 26000, Valence, France.
email: frederic.ignart@clause-vegseeds.com. Melon breeding.

Ito, Kimio. Vegetable Breeding Laboratory; Hokkaido National Agricultural Expt. Station; Hitsujigaoka, Sapporo, Japan.
email: kito@cryo.affrc.go.jp.

Jahn, Laboratory of Molly. Dept. of Plant Breeding & Genetics; 312 Bradfield Hall; Ithaca, NY 14853-1902, USA.
email: mjahn@cals.wisc.edu. Melon and squash breeding and genetics.

Johnston, Rob. Johnny's Selected Seeds; 184 Foss Hill Rd; Albion, ME 04910-9731, USA.
email: rjohnston@johnnyseeds.com. Squash and pumpkins.

Jones-Evans, Elen. Peotec Seeds SRL; Via Provinciale 42-44; 43018, Sissa (PR), Italy.
email: ejevans@peotecseeds.com.

Juarez, Benito. 37437 State Hwy 16; Woodland, CA 95695, USA.
email: Benito.juarez@seminis.com. Watermelon & melon genetics, breeding, physiology & postharvest.

Kabelka, Eileen. Dept. of Horticultural Science; 1301 Fifield Hall, Hull Road; University of Florida; Gainesville, FL 32611-0690, USA.
email: ekabelka@ifas.ufl.edu.

Karchi, Zvi. 74 Hashkedim St.; Qiryat-Tiv´on, 36501, Israel.
Cucurbit breeding, cucurbit physiology.

Kato, Kenji. Faculty of Agriculture; Okayama University; 1-1-1 Tsushima Naka; Okayama, 700-8530, Japan.
email: kenkato@cc.okayama-u.ac.jp. Use of molecular markers for QTL mapping and cultivar identification in melon.

Katzir, Nurit. Newe Ya'ar Research Center, ARO; P.O. Box 1021; Ramat Yishay, 30095 , Israel.
email: katzirn@volcani.agri.gov.il.

Kenigswald, Merav. Hazera Genetics; Mirhor M.P. Lachish; Darom, 79354, Israel.
email: meravk@hazera.com.

King, Stephen R. Vegetable & Fruit Improvement Center; Dept. of Horticultural Science; Texas A&M University; College Station, TX 77843-2133, USA.
email: srking@tamu.edu. Watermelon breeding.

Kirkbride, Jr., Joseph H. U.S. National Arboretum; 3501 New York Ave. NE; ; Washington, DC 20002-1958, USA.
email: joseph.kirkbride@ars.usda.gov. Taxonomy of *Cucumis*.

Knerr, Larry D. Shamrock Seed Company; 3 Harris Place; Salinas, CA 93901-4586, USA.
email: lknerr@shamrockseed.com. Varietal development of honeydew and cantaloupe.

Kobori, Romulo Fujito. Av. Dr. Plínio Salgado, no 4320; Bairro Uberaba; CEP 12906-840; Braganca, Paulista Sao Paulo, Brazil.
email: romulo.kobori@sakata.com.br. .

Kousik, Chandrasekar (Shaker). USDA-ARS; 2700 Savannah Hwy; Charleston, SC 29414, USA.
email: shaker.kousik@ars.usda.gov.

Kraakman, Peter. DeRuiter Zohen; Torre Caribe 7D; Aguadulce, Almeria, Spain.
email: Peter.Kraakman@deruiterseeds.com.

Kumar, Rakesh. 2336 Champion Court; Raleigh, NC 27606, USA.
email: rklnu@ncsu.edu. Integration of conventional breeding with molecular techniques.

Lanini, Brenda. Harris Moran Seed Co.; 9241 Mace Blvd.; Davis, CA 95616, USA.
email: b.lanini@harrismoran.com.

Lebeda, Aleš. Faculty of Science, Dept. Botany; Palacky University; Slechtitelu 11; 783 71, Olomouc-Holice, Czech Republic.
email: ales.lebeda@upol.cz; http://botany.upol.cz. Cucurbitaceae family, genetic resources, diseases, fungal variability, resistance breeding, tissue culture.

Lee, Chiwon W.. Dept of Plant Sciences; Loftsgard Hall, NDSU; PO Box 6050; Fargo, ND 58108-6050, USA.
email: chiwon.lee@ndsu.edu.

Legnani, Robert. Takii France; Quart de le Malgue; 13630, Etragues, France.
email: r.legnani@takii.fr.

Lehmann, Louis Carl. Louie's Pumpkin Patch; Poppelvägen 6 B; SE-541 48, Skövde , Sweden.
email: louis.lehmann@pumpkinpatch.se. *Cucurbita* - testing of squash and pumpkin for use in Southern Sweden.

Lelley, Tamas. Univ. of Nat. Resources & Applied Life Sci, Dept. for AgroBiotech. IFA; Institute for Plant Prod. Biotechnology; Konrad Lorenz Str. 20; A-3430, Tulln, Austria.
email: tamas.lelley@boku.ac.at. *Cucurbita* spp.

Lester, Gene. USDA/ARS; Kika de la Garza Subtropical Agric. Research Center; 2413 E. Highway 83, Bldg. 200; Weslaco, TX 78596, USA.
email: gene.lester@ars.usda.gov. Stress, pre/post harvest physiology, human wellness nutrient content of melons.

Lin, Depei. Sichuan Acad. Agric. Science; No. 20 Jingjusi Road; Chengdu, 610066 , P.R. China.
email: pyplin@126.com. Watermelon, melon and cucurbit breeding.

Ling, Kai-shu. USDA, ARS, U.S. Vegetable Laboratory; 2700 Savannah Hwy; Charleston, SC 29414, USA.
email: kling@saa.ars.usda.gov. Breeding for viral resistance; molecular markers.

Liu, Wenge. Zhengzhou Fruit Research Inst.; Chinese Acad. of Agric. Sci.; Gangwan Rd 28, Guancheng District; Zhengzhou,, Henan 450009 , P.R. of China.
email: lwgwm@yahoo.com.cn. Watermelon breeding, male sterility, tetraploids, triploids.

Lopez-Anido, Fernando. Universidad Nacional Rosario; CC 14; Zavalla, S 2125 ZAA , Argentina.
email: felopez@fcagr.unr.edu.ar. Breeding of *Cucurbita pepo* L. (caserta type).

Lower, Richard L. Dept. of Horticulture; Univ. of Wisconsin; 1450 Linden Drive; Madison, WI 53706, USA.
email: rllower@wisc.edu. Effects of plant type genes on yield, sex-expression, growth parameters, pest resistance & adaptablility.

Loy, J. Brent. Plant Biology Dept., G42 Spaulding Hall; Univ. of New Hampshire; 38 College Rd; Durham, NH 3824, USA.
email: jbloy@cisunix.unh.edu. Squash, melon, pumpkin. Genetics, breeding, plasticulture, mulch rowcovers.

Ma, Qing. College of Plant Protection; Northwest Agri. & Forestry University; Yangling, Shaanxi, 712100 , P.R. China.
email: maquing@nwsuaf.edu.cn. Cucumber disease resistance, resistance mechanisms.

Majde, Mansour. Gautier Semences; Route de'Avignon; 13630, Eyragues, France.
email: mansour.majde@gautiersemences.com.

Maluf, Wison Roberto. Dept. de Agricultura/UFLA; Caixa Postal 3037; 37200-000, Lavras-MG, Brazil.
email: wrmaluf@ufla.br. Cucumbers, melons, squashes.

Matsumoto, Yuichi. 3-18-10-203; Okazaki; Ami-Machi, Ibaraki 300-0335, Japan.
email: yutamn@yahoo.co.jp.

Maynard, Donald N. University of Florida; Gulf Coast Res. & Edu. Center; 14625 CR 672; Witmauma, FL 33598, USA.
email: drdon1@comcast.net. Tropical moschata improvement; watermelon variety evaluation and production practices.

Mazereeuw, Jaap. ENZA ZADEN Beheer B.V.; Postbox 7; 1600 AA, Enkhuizen, Netherlands.
email: info@enzazaden.nl.

McCreight, J.D. USDA-ARS; 1636 E. Alisal St.; Salinas, CA 93905, USA.
email: jmccreight@pw.ars.usda.gov. Melon breeding and genetics.

Myers, James R. Dept. Horticulture; Oregon State University; 4037 Ag Life Sciences Bldg.; Corvalis, OR 97331-7304, USA.
email: myersja@hort.oregonstate.edu.

Neill, Amanda. The Botanical Research Inst. of Texas; 509 Pecan St.; Fort Worth, TX 76102-4060, USA.
email: aneill@brit.org. *Gurania* and *Psiguria*.

Ng, Timothy J. Dept. Natural Resource Sci.; Univ. of Maryland; College Park, MD 20742-4452, USA.
email: binkley@umd.edu; cucurbit.genetics.cooperative@gmail.com. Melon breeding and genetics; postharvest physiology; seed germination.

Niemirowicz-Szczytt, Katarzyna. Agriculture Univ.; Dept. of Plant Genetics, Breeding and Biotechnology; St. Nowoursynowska 159; 02-766, Warsaw, Poland.
email: niemirowicz@alpha.sggw.waw.pl. Winter and summer squash, watermelon, genetics, breeding, tissue culture, biotechnology.

Nuez Viñales, Fernando. Instituto de Conservación, COMAV; Univ. Politécnica; Camino de Vera s/n; 46022, Valencia, Spain.
email: fnuez@btc.upv.es. Genetics and plant breeding.

Ochi, Yasufumi. Japan Horticultural Production Research Inst. 2-5-1; Kamishiki; Matsudo, Chiba 270-2221, Japan.
email: ochi@enken.or.jp.

Om, Young-Hyun. #568-3 Pajang-Dong; Jangan-Gu; Suwon, 440-853, Republic of Korea.
email: omyh2673@hanmail.net. Breeding of cucurbit vegetables.

Ouyang, Wei. 1603 Hampton Dr.; Davis, CA 95616, USA.
email: weiouyang1@yahoo.com. Squash, watermelon, cucumber breeding.

Owens, Ken. Magnum Seeds, Inc.; 5825 Sievers Road; Dixon, CA 95620, USA.
email: kobreeding@hotmail.com. Cucumber breeding.

Pachner, Martin. BOKU-Univ. of Nat. Resources and Applied Life Sci.; Dept. for AgroBiotech, Inst. for Plt. Prod. Biotec.; Konrad Lorenz Str. 20; A-3430, Tulln, Austria.
email: pachner@ifa-tulln.ac.at.

Palomares, Gloria. Dept. Biotecnología; Univ. Politécnica; Camino de Vera, s/n; E-46022, Valencia, Spain.
email: gpaloma@btc.upv.es. Genetic improvement in horticultural plants.

Paris, Harry. Dept. Vegetable Crops; A.R.O. Newe Ya'ar Research Ctr.; P.O. Box 1021; Ramat Yishay, 30-095, Israel.
email: hsparis@volcani.agri.gov.il. Breeding and genetics of squash and pumpkin.

Park, Soon O. TexaAgriLife Res. & Extension Ctre.; Texas A & M University; 2415 East Highway 83; Weslaco, TX 78596, USA.
email: so-park@tamu.edu. melon genetics.

Peiro Abril, José Luis. Apartado de Correos no. 2; E 04720, Aguadulce, Spain.
email: peiroab@larural.es,jlp@ramiroarnedo.com. Melon, cantaloupe, watermelon, squash, cucumber breeding, gentics, in vitro.

Perl-Treves, Rafael. Faculty of Life Sciences; Bar-Ilan University; Ramat-Gan, 52900, Israel.
email: perl@brosh.cc.biu.ac.il.

Picard, Florence. Vilmorin; Route du Manoir; 49 250, La Minitre, France.
email: Florence.picard@vilmoria.com.

Pitrat, Michel. INRA; Domaine St. Maurice; BP 94; 84143, Montfavet cedex , France.
email: Michel.Pitrat@avignon.inra.fr. Melon, disease resistance, mutants, genetic map.

Polewczak, Lisa. Syngenta Seeds; 10290 Greenway Road; Naples, FL 34114, USA.
email: l.polewczak@gmail.com. Squash, cantaloupe, watermelon breeding & genetics.

Poostchi, Iraj. 97 St. Marks Road; Henley-on-Thames; The United Kingdom, RG9 1LP, United Kingdom. Breeding cantaloupes, melons and watermelons.

Portyankin, Aleksey. Research Institute of Greenhouse Veg. Prod.; 2-Khutorskaya 11/1; Stroenie 1; 127287, Moscow, Russia.
email: port75_alex@mail.ru.

Poulos, Jean M. Nunhems USA, Inc.; 7087 E. Peltier Rd.; Acampo, CA 95220, USA.
email: jean.poulos@nunhems.com. Melon breeding.

Randhawa, Lakhwinder. Sakata Seed America, Inc.; 2854 Niagara Ave; Colusa, CA 95932, USA.
email: lrandhawa@sakata.com. Molecular markers.

Randhawa, Parm. CA Seed & Plant Lab; 7877 Pleasant Grove Rd.; Elverta, CA 95626, USA.
email: randhawa@calspl.com.

Ranganath, Srinivas. #176, 6th Cross, HMT Layout; Mathikere, Bangalore 560054, India.
email: prososrini@yahoo.com.in, s.ranganath@nunhems.com.

Ray, Dennis. Dept. Plant Sci.; Univ. of Arizona; P.O. Box 210036; Tucson, AZ 85721-0036, USA.
email: dtray@email.arizona.edu. Genetics and cytogenetics of *Cucumis melo* and Citrullus spp.

Reitsma, c/o Kathy. North Central Regional Plant Introduction Sta.; Iowa State University; Ames, IA 50011-1170, USA.
email: kathleen.reitsma@ars.usda.gov, kreitsma@iastate.edu. Curator of cucurbit germplasm.

Reuling, Gerhard T.M. Nunhems Netherlands B.V.; P.O. Box 4005; 6080, AA Haelen, Netherlands.
email: g.reuling@nunhems.com. Breeding long cucumber.

Robinson, R. W. Emeritus Prof. , Dept. Hort. Sci.; NY Agri. Expt. Station; Cornell University; Geneva, NY 14456-0462, USA.
email: rwr1@nysaes.cornell.edu. Breeding and genetics of cucurbits.

Rodenburg, Marinus. East-West Seed Indonesia; Desabenteng, Campaka; P.O. Box 1, Purwakarta, Indonesia.
email: rien@ewsi.co.id.

Rokhman, Fatkhu. PT East West Seed Indonesia; P.O. Box 1, Campaka; Purwakarta 41181, W. Java, Indonesia. ***email:*** fatkhu_Rokhman@ewsi.co.id. Cucumber, watermelon and melon breeding.

Shetty, Nischit V. Seminis Vegetable Seeds; 432 TyTy Omega Road; Tifton, GA 31794, USA.
email: nischit.shetty@seminis.com. Cucumber breeding.

Shimamoto, Ikuhiro. 146-11 Daigo; Kashihara, NARA 634-0072, Japan.
email: shimamoto@suika-net.co.jp.

Simon, Phillip W. USDA-ARS-Vegetable Crops; Dept. of Horticulture, Univ. of Wis.; 1575 Linden Dr.; Madison, WI 53706-1590, USA.
email: psimon@wisc.edu. Breeding and genetics.

Stephenson, Andrew G. 208 Mueller Lab; Penn State Univ.; University Park, PA 16802-5301, USA.
email: as4@psu.edu.

Swanepoel, Cobus. Pannar; P.O. Box 13339; Northmead, 1511, South Africa.
email: cobus.swanepoel@pannar.co.za.

Tatlioglu, Turan. Martar Tohumculuk A.S.; Han Mahallesi; 5 Eylül Caddesi No. 7; Susurluk, (Balikesir) 10600, Turkey.
email: turantatlioglu@yahoo.com. Hybrid breeding, sex inheritanc, sex genes.

Theurer, Christoph. GlaxoSmithKline Consumer Healthcare GmbH & Co. KG; Consumer Healthcare Gmbh&Co.KG; Benzstrasse 25; D-71083, Herrenberg , Germany.
email: Christoph.Theurer@gsk.com.

Thro, Ann Marie. USDA-CSREES; 1400 Independence Avenue SW, Stop 2201; Washington, DC 20260-2201, USA.
email: athro@csrees.usda.gov. USDA CSREES National Program Leader Plant Breeding, Genetics & Genomics.

Tolla, Greg. Seminis Vegetable Seeds; 37437 State Hwy 16; Woodland, CA 95695, USA.
email: greg.tolla@seminis.com. Breeding and genetics.

Vadra Halli, Satish. #1/2, Krishna Road; Basavanagudi; Bangalore, 560004 Karnataka, India.
email: satishvadrahalli@yahoo.com.

Vardi, Eyal. Origene Seeds Ltd.;Givat Brenner, 60948, Israel.
email: eyal@origeneseeds.com.

Wehner, Todd. Dept. Horticultural Science; Box 7609; North Carolina State Univ.; Raleigh, NC 95616, USA.
email: todd_wehner@ncsu.edu. Pickling/slicing cucumber, watermelon, luffa gourd, selection, disease resistance, yield, genetics and breeding.

Weng, Yiqun. USDA Vegetable Crops Research Unit; University of Wisconsin; 1575 Linden Dr.; Madison, WI 53706, USA.
email: yiqun.weng@ars.usda.gov.

Wessel-Beaver, Linda. Dept. Crops & Agroenvironemental Sciences; Call Box 9000; Univ. of Puerto Rico; Mayaguez, PR 00681-9000, USA.
email: llindawessesl.beaver@upr.edu; Lwesselbeaver @yahoo.com. Pumpkin and squash breeding and genetics; disease and insect resistance; cucurbit evolution and domestication.

Williams, Tom V. 2329 Pinewood Circle; Naples, FL 34105, USA.
email: tvwilli@earthlink.net; tvwili@aol.com. Watermelon consultant.

Winkler, Johanna. Saatzucht Gleisdorf GmbH; Am Tieberhof 33; A-8200, Gleisdorf , Austria.
email: winkler.szgleisdorf@utanet.at.

Wu, Mingzhu. Hort. Instit.; Xinjiang Acad. Agric. Sci; Nachang Road No. 38; Urumqi, Xinjiang 830000 , P.R. of China.
email: mzwu@x263.net.

Yorty, Paul. Qualiveg Seed Production; 3033 E., 3400 N.; Twin Falls, ID 83301, USA.
Cucurbit breeding.

Zhang, Shengping. Institute of Vegetables & Flowers; Chinese Academy of Agric. Sci.; No 12 Zhongguancun Nandajie; Beijing, 100081, P.R. of China.
email: zhangshp2007@hotmail.com. Cucumber genetics & breeding.

Zhang, Xingping. Syngenta Seeds; 21435 Co. Rd. 98; Woodland, CA 95695, USA.
email: xingping.zhang@syngenta.com. Watermelon and melon genetics & breeding.

Zhou, Ihichen. Xinjiang XiYu Seeds Co., LTD.; No. 32 Eastern Ningbian Road; Changji, Xingiang 831100, P.R. of China.
email: xiyuseedsxj@hotmail.com. Breeding watermelon, melon, squash and other cucurbits.

2008-2009 CGC Membership by Country

Argentina
Lopez-Anido, Fernando

Australia
Herrington, Mark

Austria
Lelley, Tamas
Pachner, Martin
Winkler, Johanna

Brazil
Gardingo, Jose Raulindo
Kobori, Romulo Fujito
Maluf, Wison Roberto

China, Peoples Republic of
Bao, HaiQing
Chen, Jin
Feng Haizhen, Li
Lin, Depei
Ma, Qing
Liu, Wenge
Wu, Mingzhu
Zhang, Shengping
Zhou, Ihichen

Czech Republic
Holman, Bohuslav
Lebeda, Aleš

Egypt
Aboul-Nasr, M. Hossam

France
Baudracco-Arnas,, Sylvie
Bertrand, Francois
Boissot, Nathalie
Ignart, Frederic
Legnani, Robert
Majde, Mansour
Picard, Florence
Pitrat, Michel

Germany
Theurer, Christoph

India
Aurangabadkar, Laxman
Behera, Tusar
Ranganath, Srinivas
Vadra Halli, Satish

Indonesia
Rodenburg, Marinus
Rokhman, Fatkhu

Israel
Belotserkovsky, Harel
Burger, Yosi
Cohen, Roni
Davidi, Haim
Karchi, Zvi
Katzir, Nurit
Kenigswald, Merav
Paris, Harry
Perl-Treves, Rafael
Vardi, Eyal

Italy
de Groot, Erik
Ficcadenti, Nadia
Gatto, Gianni
Jones-Evans, Elen

Japan
Furuki, Toshi
Hagihara, Toshitsugu
Ito, Kimio
Kato, Kenji
Matsumoto, Yuichi
Ochi, Yasufumi
Shimamoto, Ikuhiro

Korea, Republic of
Cho, Myeong-Cheoul
Om, Young-Hyun

Malta
Attard, Everaldo

Mexico
Garza-Ortega, Sergio

Netherlands, The
Bal, Eric
De Ruiter, Wouter
Hertogh, Kees
Hofstede, Rene
Hoogland, Jan
Mazereeuw, Jaap
Reuling, Gerhard T.M.

New Zealand
Grant, Doug

Philippines
Beronilla, Renita

Poland
Niemirowicz-Szczytt, Katarzyna

Russia
Portyankin, Aleksey

Serbia
Berenji, Janos

South Africa
Swanepoel, Cobus

Spain
Abad Martin, Jesus
Buil Benedi, María Angeles
Deleu, Wim
Den Hertog, Maarten
Gómez-Guillamón, Maria L.
Kraakman, Peter
Nuez Viñales, Fernando
Palomares, Gloria
Peiro Abril, José Luis
Lehmann, Louis Carl

Thailand
Chuanchai, Vinich
de Hoop, Simon Jan
Duangsong, Usa
Turkey
Tatlioglu, Turan

United Kingdom
Dawson, Halina
Poostchi, Iraj

USA
Andres, Thomas C.
Bell, Duane
Block, Charles C.
Boyhan, George E.
Brown, Rebecca
Connolly, Bryan
Crosby, Kevin

Dane, Fenny
Davis, Angela
Dombrowski, Cory
Everts, Kathryne
Frobish, Mark
Froese, David C.
Gabor, Brad
Goldman, Amy P.
Groff, David
Grumet, Rebecca
Guner, Nihat
Gusmini, Gabriele
Havey, Michael J.
Himmel, Phyllis
Huan, Jin
Jahn, Molly
Johnston, Rob
Juarez, Benito
Kabelka, Eileen
King, Stephen R.
Kirkbride, Jr., Joseph H.
Knerr, Larry D.
Kousik, Chandrasekar (Shaker)
Kumar, Rakesh
Lanini, Brenda
Lee, Chiwon W.
Lester, Gene
Ling, Kai-shu
Lower, Richard L.
Loy, J. Brent
Maynard, Donald N.
McCreight, J.D.
Myers, James R.
Neill, Amanda
Ng, Timothy J.
Ouyang, Wei
Owens, Ken
Park, Soon O.
Polewczak, Lisa
Poulos, Jean M.
Randhawa, Lakhwinder
Randhawa, Parm
Ray, Dennis
Reitsma, Kathy
Robinson, R. W.
Shetty, Nischit V.
Simon, Phillip W.
Stephenson, Andrew G.
Thro, Ann Marie
Tolla, Greg
Wehner, Todd
Weng, Yiqun
Wessel-Beaver, Linda
Williams, Tom V.
Yorty, Paul
Zhang, Xingping

2008-2009 United States CGC Membership By State

Alabama
Dane, Fenny

Arizona
Ray, Dennis

California
Gabor, Brad
Himmel, Phyllis
Huan, Jin
Juarez, Benito
Knerr, Larry D.
Lanini, Brenda
McCreight, J.D.
Ouyang, Wei
Owens, Ken
Poulos, Jean M.
Randhawa, Lakhwinder
Randhawa, Parm
Tolla, Greg
Zhang, Xingping

Colorado
Froese, David C.

Connecticut
Connolly, Bryan

District of Columbia
Kirkbride, Jr., Joseph H.
Thro, Ann Marie

Florida
Dombrowski, Cory
Guner, Nihat
Gusmini, Gabriele
Kabelka, Eileen
Maynard, Donald N.
Polewczak, Lisa
Williams, Tom V.

Georgia
Boyhan, George E.
Groff, David
Shetty, Nischit V.

Iowa
Reitsma, Kathy
Block, Charles C.

Idaho
Yorty, Paul

Illinois
Frobish, Mark

Maryland
Everts, Kathryne
Ng, Timothy J.

Maine
Johnston, Rob

Michigan
Grumet, Rebecca

North Carolina
Kumar, Rakesh
Wehner, Todd

North Dakota
Lee, Chiwon W.

New Hampshire
Loy, J. Brent

New York
Andres, Thomas C.
Goldman, Amy P.
Robinson, R. W.

Ohio
Bell, Duane

Oklahoma
Davis, Angela

Oregon
Myers, James R.

Pennsylvania
Stephenson, Andrew G.

Puerto Rico
Wessel-Beaver, Linda

Rhode Island
Brown, Rebecca

South Carolina
Kousik, Chandrasekar (Shaker)
Ling, Kai-shu

Texas
Crosby, Kevin
King, Stephen R.
Lester, Gene
Neill, Amanda
Park, Soon O.

Wisconsin
Havey, Michael J.
Lower, Richard L.
Simon, Phillip W.
Weng, Yiqun

Covenant and By-Laws of the Cucurbit Genetics Cooperative

ARTICLE I. Organization and Purposes

The Cucurbit Genetics Cooperative is an informal, unincorporated scientific society (hereinafter designated "CGC") organized without capital stock and intended not for business or profit but for the advancement of science and education in the field of genetics of cucurbits (Family: Cucurbitaceae). Its purposes include the following: to serve as a clearing house for scientists of the world interested in the genetics and breeding of cucurbits, to serve as a medium of exchange for information and materials of mutual interest, to assist in the publication of studies in the aforementioned field, and to accept and administer funds for the purposes indicated.

ARTICLE II. Membership and Dues

1. The membership of the CGC shall consist solely of active members; an active member is defined as any person who is actively interested in genetics and breeding of cucurbits and who pays biennial dues. Memberships are arranged by correspondence with the Chairman of the Coordinating Committee.
2. The amount of biennial dues shall be proposed by the Coordinating Committee and fixed, subject to approval at the Annual Meeting of the CGC. The amount of biennial dues shall remain constant until such time that the Coordinating Committee estimates that a change is necessary in order to compensate for a fund balance deemed excessive or inadequate to meet costs of the CGC.
3. Members who fail to pay their current biennial dues within the first six months of the biennium are dropped from active membership. Such members may be reinstated upon payment of the respective dues.

ARTICLE III. Committees

1. The Coordinating Committee shall govern policies and activities of the CGC. It shall consist of six members elected in order to represent areas of interest and importance in the field. The Coordinating Committee shall select its Chairman, who shall serve as a spokesman of the CGC, as well as its Secretary and Treasurer.
2. The Gene List Committee, consisting of at least five members, shall be responsible for formulating rules regulating the naming and symbolizing of genes, chromosomal alterations, or other hereditary modifications of the cucurbits. It shall record all newly reported mutations and periodically report lists of them in the Report of the CGC. It shall keep a record of all information pertaining to cucurbit linkages and periodically issue revised linkage maps in the Report of the CGC. Each committee member shall be responsible for genes and linkages of one of the following groups: cucumber, *Cucurbita* spp., muskmelon, watermelon, and other genera and species.
3. Other committees may be selected by the Coordinating Committee as the need for fulfilling other functions arises.

ARTICLE IV. Election and Appointment of Committees

1. The Chairman will serve an indefinite term while other members of the Coordinating Committee shall be elected for ten-year terms, replacement of a single retiring member taking place every other year. Election of a new member shall take place as follows: A Nominating Committee of three members shall be appointed by the Coordinating Committee. The aforesaid Nominating Committee shall nominate candidates for an anticipated opening on the Coordinating Committee, the number of nominees being at their discretion. The nominations shall be announced and election held by open ballot at the Annual Meeting of the CGC. The nominee receiving the highest number of votes shall be declared elected. The newly elected member shall take office immediately.
2. In the event of death or retirement of a member of the Coordinating Committee before the expiration of his/her term, he/she shall be replaced by an appointee of the Coordinating Committee.
3. Members of other committees shall be appointed by the Coordinating Committee.

ARTICLE V. Publications

1. One of the primary functions of the CGC shall be to issue an Annual Report each year. The Annual Report shall contain sections in which research results and information concerning the exchange of stocks can be published. It shall also contain the annual financial statement. Revised membership lists and

other useful information shall be issued periodically. The Editor shall be appointed by the Coordinating Committee and shall retain office for as many years as the Coordinating Committee deems appropriate.

2. Payment of biennial dues shall entitle each member to a copy of the Annual Report, newsletters, and any other duplicated information intended for distribution to the membership. The aforementioned publications shall not be sent to members who are in arrears in the payment of dues. Back numbers of the Annual Report, available for at least the most recent five years, shall be sold to active members at a rate determined by the Coordinating Committee.

ARTICLE VI. Meetings

An Annual Meeting shall be held at such time and place as determined by the Coordinating Committee. Members shall be notified of time and place of meetings by notices in the Annual Report or by notices mailed not less than one month prior to the meeting. A financial report and information on enrollment of members shall be presented at the Annual Meeting. Other business of the Annual Meeting may include topics of agenda selected by the Coordinating Committee or any items that members may wish to present.

ARTICLE VII. Fiscal Year

The fiscal year of the CGC shall end on December 31.

ARTICLE VIII. Amendments

These By-Laws may be amended by simple majority of members voting by mail ballot, provided a copy of the proposed amendments has been mailed to all the active members of the CGC at least one month previous to the balloting deadline.

ARTICLE IX. General Prohibitions

Notwithstanding any provisions of the By-Laws or any document that might be susceptible to a contrary interpretation:

1. The CGC shall be organized and operated exclusively for scientific and educational purposes.
2. No part of the net earnings of the CGC shall or may under any circumstances inure to the benefit of any individual.
3. No part of the activities of the CGC shall consist of carrying on propaganda or otherwise attempting to influence legislation of any political unit.
4. The CGC shall not participate in, or intervene in (including the publishing or distribution of statements), any political campaign on behalf of a candidate for public office.
5. The CGC shall not be organized or operated for profit.
6. The CGC shall not:
   a. lend any part of its income or corpus without the receipt of adequate security and a reasonable rate of interest to;
   b. pay any compensation in excess of a reasonable allowance for salaries or other compensation for personal services rendered to;
   c. make any part of its services available on a preferential basis to;
   d. make any purchase of securities or any other property, for more than adequate consideration in money's worth from;
   e. sell any securities or other property for less than adequate consideration in money or money's worth; or
   f. engage in any other transactions which result in a substantial diversion of income or corpus to any officer, member of the Coordinating Committee, or substantial contributor to the CGC.

The prohibitions contained in this subsection (6) do not mean to imply that the CGC may make such loans, payments, sales, or purchases to anyone else, unless authority be given or implied by other provisions of the By- Laws.

ARTICLE X. Distribution on Dissolution

Upon dissolution of the CGC, the Coordinating Committee shall distribute the assets and accrued income to one or more scientific organizations as determined by the Committee, but which organization or organizations shall meet the limitations prescribed in sections 1-6 of Article IX.

5523211R0

Made in the USA
Charleston, SC
27 June 2010